Lecture Notes in Earth Sciences

Edited by Somdev Bhattacharji, Gerald M. Friedman,
Horst J. Neugebauer and Adolf Seilacher

5

Paleogeothermics

Evaluation of Geothermal Conditions in the Geological Past

Edited by Günter Buntebarth and Lajos Stegena

Springer-Verlag

Berlin Heidelberg New York London Paris Tokyo

Editors

Dr. Günter Buntebarth
Technische Universität Clausthal, Institut für Geophysik
Arnold-Sommerfeld-Str. 1, D-3392 Clausthal-Zellerfeld, FRG

Prof. Dr. Lajos Stegena
Institute of Environmental Physics, Eötvös-University
Kun Béla Tér 2, H-1083 Budapest, Hungary

ISBN 3-540-16645-9 Springer-Verlag Berlin Heidelberg New York
ISBN 0-387-16645-9 Springer-Verlag New York Heidelberg Berlin

This work is subject to copyright. All rights are reserved, whether the whole or part of the material is concerned, specifically those of translation, reprinting, re-use of illustrations, broadcasting, reproduction by photocopying machine or similar means, and storage in data banks. Under § 54 of the German Copyright Law where copies are made for other than private use, a fee is payable to "Verwertungsgesellschaft Wort", Munich.

© Springer-Verlag Berlin Heidelberg 1986
Printed in Germany

Printing and binding: Beltz Offsetdruck, Hemsbach/Bergstr.
2132/3140-543210

PREFACE

During the last decades, remarkable progress in heat flow studies has been made and a rough picture of the global surface heat flow density distribution can now be drawn. Simultaneously, the question of over which time period the surface heat flow is constant arose.

There is a big field of model calculations, based on the changes in radioactive heat generation of the Earth, on plate motions, on stretching hypotheses or on other ideas, which result in geotherms in the geological past. Although these speculative paleogeotherms seem to be realistic especially in oceanic areas they do not belong to the scope of this book. In continental areas however, it is not possible to find a simple time dependence of the surface heat flow density. However, petroleum research and tectogenetic studies are very interested in the geothermal history of sedimentary basins and other continental areas. To obtain satisfactory results, a more or less direct determination of paleo heat flow density or geothermal gradient would be necessary to give more certain boundary conditions for calculating oil generation, and for controlling tectogenetic hypotheses.

There are many methods available in the geosciences to determine temperatures in the geological past. Most of these models are able to estimate temperatures at which a mineral or a mineral assemblage was formed. These methods, however, are mostly unsuitable to reach the main goal of paleogeothermics in general, which is to determine the (regional) heat flow density variations during the geological past for bigger geological units, such as sedimentary basins.

The methods applied most in sedimentary basins have been deduced from the degree of coalification of organic matter. Although much effort has been made to explain analytically the organic metamorphism, the results found up to now have been insufficient. However, the widespread application of this thermometer to estimate ancient thermal conditions is also reflected in the contents of this very volume where the interpretation of the degree of coalification of organic matter plays an important role.

As well as this geothermometers, other methods are reviewed from a geophysical viewpoint which favours methods suitable to determine a paleothermal state of the upper crust.

Further contributions of this book deal with

- the history of the earth's surface temperature whose change provides an essential correction factor in heat flow density determinations,

- isotope geothermometers and their application to various environments to evaluate thermal conditions in the past geological history,

- an application of the radiometric dating method to retrace the paleothermal condition of the Central Alps.

Most of the contributions were presented at the symposium "Paleogeothermics" which was held at the 18. General Assembly of the International Union of Geodesy and Geophysics, August 15-27, 1983 in Hamburg/FRG.

It has been the first time that such a symposium has been organized by the International Heat Flow Commission, and this book presents an attempt to define paleogeothermics under the auspices of the International Heat Flow Commission.

G. Buntebarth
Institute of Geophysics
Technical University Clausthal

L. Stegena
Institute of Environmental Physics
Eötvös University Budapest

CONTENTS

Preface		1
Contents		3
1.	Methods in Paleogeothermics BUNTEBARTH/STEGENA	5
2.	Temperature history of the earth's surface in relation to heat flow SHACKLETON	41
3.	Isotope geothermometers HOEFS	45
4.	Relations between coalification and paleogeothermics in Variscan and Alpidic foredeeps of western Europe TEICHMÜLLER/TEICHMÜLLER	53
5.	The correlation of vitrinite reflectance with maximum temperature in humic organic matter BARKER/PAWLEWICZ	79
6.	A comparison of two vitrinite reflectance methods for estimating paleotemperature gradients BUNTEBARTH/MIDDLETON	95
7.	Methods for paleotemperature estimating using vitrinite reflectance data: a critical evaluation VETŐ/DÖVENY	105
8.	A reaction kinetic approach to the temperature-time history of sedimentary basins SAJGO/LEFLER	119
9.	Limits of application of the reaction kinetic method in paleogeothermics LEFLER/SAJGO	153
10.	Geothermal effect of magmatism and its contribution to the maturation of organic matter in sedimentary basins HORVATH/DÖVENY/LACZO	173
11.	Paleotemperatures in the Central Alps - an attempt of interpretation WERNER	185
12.	Geothermal studies in oil field districts of north China WANG/WANG/YAN/LU	195
References		205
Subject Index		229

METHODS IN PALEOGEOTHERMICS

BUNTEBARTH, G.* and L. STEGENA**

* Institut für Geophysik, TU Clausthal
Arnold-Sommerfeld-Str. 1, D-3392 Clausthal-Zellerfeld, F.R. of Germany

** Institute of Geophysics, Eötvös-University
Kun Béla Tér 2, H-1083 Budapest

Introduction

An attempt is made to bring together geophysical, geological and geochemical methods bearing on ancient thermal conditions of the earth's crust. Methods are emphasized which are suitable to estimate temperature gradients in the past, in order to evaluate the evolution of or merely the changes in the thermal regime within the crust.

The application of the degree of coalification of organic matter has received particular attention as a means of estimating the geothermal history of sedimentary basins because the degree of coalification is mainly influenced by the temperature of the environment and the time of exposure at this temperature. Several empirical interpretation methods are reported which have been developed for specific basins and which are especially valid for these areas.

During crystal growth, liquids and other phases can be entrapped in the host crystal. These entrapped phases preserve the temperature and the pressure conditions which were present at the time of crystal growth.
Chemical reactions are temperature sensitive. Therefore, solution equilibria and isotope exchange reactions are applied to estimate paleothermal conditions, or to compare the calculated reaction temperature with the present thermal state in particular areas.

A recent successfully tested method which deals with the transformation of minerals during diagenesis is reported. Clay minerals, zeolites and quartz polymorphs are transformed in sedimentary rocks of similar composition at distinct temperatures.

Another method is reported which analyses the color alteration of conodonts. This method is applicable for sedimentary rocks from the Late Cambrian to the Triassic period when the conodonts lived.

Radiometric dating is the only method which yields a thermal history of crystalline rocks. Because each radioactive system has its own closure temperature, radiometric age determinations give the ages at which a rock cooled down to the respective closure temperature.

1. Diagenesis of organic matter

Since organic life grows on the earth, it is included in the geological cycle. The remains of the organic matter are covered by sediments or deposited together with clastic detritus. If the circumstances are favourable, the organic matter is preserved and subsides within a sedimentary basin. During subsidence it undergoes increasing pressure as well as temperature, and both gradually alter the original material. The alteration of organic matter is known as diagenesis and process of coalification. There are two factors which govern predominantly the rank of coalification, which are the temperature in the depth where the organic matter existed during its history, and the time of its exposure. An interpretation of the degree of coalification based on the effect of temperature and time of exposure to that temperature, can be of limited value only. More care must be taken on the origin of organic material and the first steps in its structural and chemical decomposition in different environments.

The oldest coals which seem to be of plant origin are preserved in rocks of Algonkium age in North America. Several localities with coal embedded in a sedimentary sequence are known in the Lower Devonian. Since Middle and Upper Devonian, when plenty of plants grew on the continent and on the submerged shore, coal seams are more common. The most prominent bituminous coal deposits are of Carboniferous age in Europe and North America, and of Permian, Triassic and Jurassic age in South Africa, Eastern Australia and India.

Since Cretaceous, much more variety in the flora has been created which implies more heterogeneity in the plant remains from which the coaly matter originates. The coals are formed not only from different plant communities but also at different environmental conditions which are summarized by M. TEICHMÜLLER & R. TEICHMÜLLER (1981). It is important, that the plants or their remains have to be deposited under conditions with restricted oxygen supply. Usually, this condition is present in swampy areas. If a sedimentary basin with swampy areas subsides gradually, the organic matter can be deposited in layers of some thickness. A warm or temperate to cool climate with high humidity throughout the year is necessary to retain the condition favoured for organic deposition.

There are a few peat-forming plant communities which grow in different swamp types, i.e. moss swamp, forest swamp, open reed swamps and partly submerged areas with water plants. The most productive areas are forest swamps under tropical conditions. Economical coal seams yield from deposition in swamps, in general. As well as in coal seams, organic matter is also present in a dispersed form in many minerogene sedimentary rocks. Plant remains in river deltas and on the shores of lakes and oceans, barks, other plant detritus, and also coal which can be redeposited, can be covered by clastic sediments and buried. If the environmental conditions are favourable for preservation, the organic substances undergo the diagenesis during the subsidence, and will become coaly particles like the plant remains in swamps. However, there is a difference. The plant remains are exposed not only to the mechanical treatment during the transport by water, but also to the oxidizing atmosphere and to the bacterial activity at the surface which favours the preservation of especially resistant particles. This means that the original organic substance is not exactly the same as in seams. The composition of the organic matter bearing rocks is of some importance too. The organic matter is often oxidized in sandstones, especially in red-coloured ones, but is rather seldom in limestone. Usually clay and siltstones are the rock types from which the organic particles can be observed and interpreted for paleogeothermal investigations.

There are additional factors influencing the composition of the organic substance which yields the coaly matter. Whereas organic deposits under terrestrial and subaquatic conditions are comparable, marine-influenced and calcium-rich swamps produce coals richer in ash, sulphur and nitrogen. These conditions imply that a different acidity of water may produce coals of same distinguishable properties, even with the same original material. It seems that the bacterial activity is a most important factor controlling the decomposition of plants and thereby at least the original materials for the coals. Therefore, all environmental properties which favour or prevent bacterial life also define the properties of the coal. Vitrinite is a most common coalification product which is formed from organic deposits under some acid condition. If the environment is neutral to weakly alkaline, the bacterial activity is very high. Since the protein of the bacteria is also accumulated, the organic substances yield hydrogen-rich bituminous products which form bituminite and weakly reflecting vitrinites during subsidence (M. TEICHMÜLLER & R. TEICHMÜLLER, 1981).

Peat is the first stage in the diagenetic process of the organic matter. Peatification can start after the burial of plant remains with the help of the bacteria, which are active down to some meters of depth. With continuing subsidence, the increasing overburden pressure causes the water to be squeezed out of the organic substances. The temperature during this physical process may range between about 20 to 50° C. At the upper limit of the temperature range, little methane is split off (van HEEK et al., 1971), and the transformation from peat to brown coal is usually reached in a depth range between 200 m and 400 m. At temperatures of about 70 to

100° C CO_2 is released, and at temperatures of about 160 to 200° C, at which low volatile bituminous coal gradually changes to semi-anthracite, large quantities of methane develop.

The rank of coal is determined in a general way by appearances and/or by its properties, e.g. bright brown coal and gas coal. This qualitative scale is not sufficient for analytical investigations. The composition of organic matter in sediments is 90 % kerogen and 10 % bitumen (hydrocarbon, resin, asphaltene). The fraction soluble in organic solvents, is called bitumen, whereas the other fraction, insoluble in organic matter, is termed kerogen. There are methods to estimate the maturity by examining the soluble organic matter: percentage carbon in bitumen, carbon preference index (odd carbon number compounds to even carbon number), paraffin profile, percentage wet gas. Other, more important methods, examine the kerogen as a maturation index. These methods are the kerogen alteration index KAI, thermal alteration index TAI, pyrolysis, elementary CHO analysis, and atomic H/C ratio.

All these chemical rank parameters are not applicable in general for rocks with finely dispersed organic matter, because the chemical methods need some amount of organic particles. The rank determination with microscope is successful. The method is not destructive for the sample, and is easy to apply. Vitrinite is the most common coal maceral, and is the one taken in order to measure its optical reflectivity at the polished sample under oil, applying monochromatic light. This method is applicable to both the coal from seams and the coaly particles dispersed in sedimentary rocks.

Vitrinite reflectivity is a ratio of the intensity of the reflected light and the source light, expressed in percent, using vitrinite (= woody kerogen) as the reflector. The value is often simply called R_o, % R_o, or % Rm the subscript "o" designates that the measurement was made in oil, and "m" means the mean reflectivity, instead of Rmax, the maximum reflectivity, which should be applied at reflectivity values above 4 % Rm.

The reflectivity coefficient gives a continuous scale for the coalification of huminite/vitrinite with values ranging from about 0.2 % up to more than 5 % (M. TEICHMÜLLER, 1970). Huminite and vitrinite are maceral groups of humous components, where huminite is the precursor of vitrinite in peat and brown coal. During the progress in coalification huminite is converted into vitrinite between the coalification stages of dull and bright brown coal.

If some rocks are so poor in organic matter that concentrates must be prepared by chemical or physical methods, it is much more difficult to determine the correct degree of coalification. The surroundings of the particles are often helpful to select the representative ones for measurement. The selection of the correct coal macerals, i.e. vitrinite, poses the greatest difficulty in the determination of the degree of coalification in rocks. For this determination the so-called "kerobitumen" which can be found in bituminous shales is of some importance. The bituminous matter reflects in the lower rank of coalification less than vitrinite, but more in the rank of anthracite. The distinction between recycled and authochtonous organic matter is often difficult in rocks, but nearly impossible in concentrates. There are a lot of problems arising from the selection of macerals for measurements, which are described more detailed e.g. in STACH et al. (1982), ROBERT (1985), TISSOT & WELTE (1978).

Besides the reflectivity of vitrinite in shales, sandstones and limestones with dispersed coaly particles, the spectral fluorescence measurements on sporinite has been introduced as an indicator of the degree of diagenesis. If sporinite is irradiated with ultraviolet light ($\lambda = 365 \pm 30$ nm), a visible fluorescence can be observed from yellow to dark red colour. However, the sporinite fluorescence spectra are observed at low grades of diagenesis only, i.e. from the stage of peat to that of high volatile bituminous coal (OTTENJANN et al., 1974).

Both parameters, the reflectivity of vitrinite and the sporinite fluorescence, are used together to find a more correct degree of diagenesis. The interpretation of the rank of coalification for paleogeothermics is based on the fact that the temperature is the most important factor that increases the degree of coalification, but the duration of heating must also be taken into consideration. The influence of pressure, however, seems to be negligible. Based on HUCK & KARWEIL (1955), LOPATIN (1971) gave a simple scheme for describing the degree of coalification. Supposing that the coalification process is to be treated as a first order chemical reaction, the Arrhenius' equation is valid and the velocity of the reaction (k) depends exponentially on temperature:

$$k = a \exp(-E/RT)$$

(a: frequency-factor, E: activation energy, R: gas-constant, T: temperature in Kelvin). Numerous chemical reactions double their reaction velocity for each 10° C temperature growth, not far from room temperatures, because their activation energy lies around 54 kJ/mole.

LOPATIN (1971) accepted this value and suggested that the dependence of maturity on time is linear, and the dependence on temperature has an exponential character. Therefore, the velocity of the "coalification" reaction can be written as

$$k \sim 2^{0.1T(t)}$$

and the parameter which describes the rank of coalification

$$C \sim \int_0^{t^*} 2^{0.1T(t)} dt$$

where $T(t)$ is the temperature of the layer during the time interval dt, and t^* is the time from the deposition of the layer till the present.

For practical reasons, LOPATIN introduced the sum instead of the integral, dividing the whole temperature history of the layer into 10° C temperature intervals. He then arbitrarily chose the 100 to 110° C temperature interval (which is the mean domain of oil generation) as the base interval and assigned to it an index value of $n = 0$, to the 120 - 130° C interval $n = 2$, to the 90 - 100° C interval $n = -1$, and so on. The maturity parameter calculated in this manner was called the Time Temperature Index (TTI),

$$TTI = \sum_{nmax}^{nmin} (\Delta t_n) 2^n$$

where Δt_n is the time interval (in Ma) the layer spent in the n-th 10° C temperature interval, and nmax and nmin are the n-values of the highest and lowest temperature intervals occurring in the thermal history of the layer.

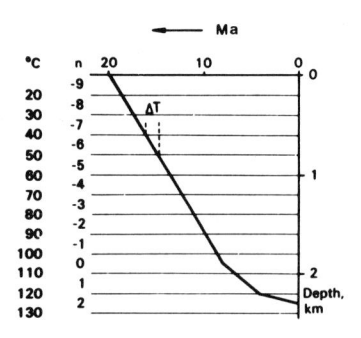

Fig. 1. LOPATIN's (1971) method for the calculation of the Time Temperature Index for a layer lying at a depth of 2300 m, aged 20 Ma. TTI is characteristic for the maturity of organic matter.

Fig. 1 demonstrates the method of calculation of TTI, for a hypothetical layer 20 Ma old and lying at present at a depth of 2300 m. Let us suppose that the subsidence and burial history of the layer during geologic time was determined as shown by the curve of Fig. 1. Let us then suppose that the present geothermal gradient is 50 mK/m, and the gradient was constant during the whole sedimentary history, as shown in Fig. 1, by the horizontal straight geotherms. In this case for the layer of Fig. 1, TTI = 15.2.

Based on 402 thermal maturity (R_o) data from 31 worldwide wells, WAPLES (1980) determined a correlation between TTI values calculated for each borehole from burial histories, supposing the validity of present geothermal conditions during the geological past, and R_o values measured (Fig. 2).

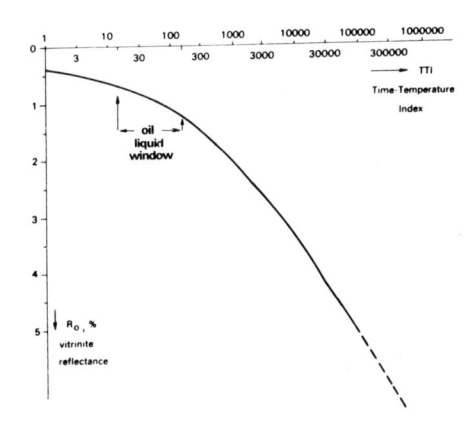

Fig. 2. Correlation between the Time Temperature Index of maturity and vitrinite reflectance R_o (after WAPLES, 1980)

These antecedents make possible the paleo heat flow estimation for a borehole, by the following steps:

- Based on known ages of some sedimentary layers in the borehole, the sedimentary history for these layers is determined (Fig. 3, dotted lines).

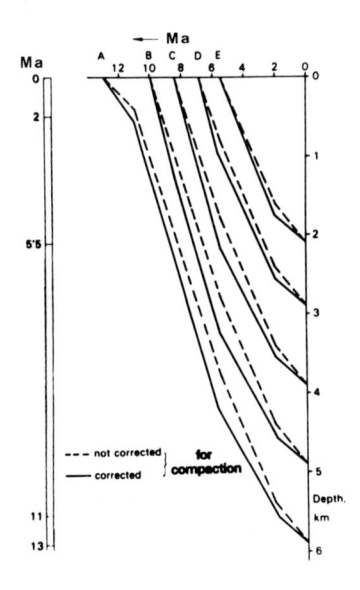

Fig. 3. Sedimentary history of a borehole (HOD) in the Pannonian basin, calculated on the basis of the ages in the left hand column, with and without correction of compaction (STEGENA et al., 1981)

- Using porosity-depth functions and/or other considerations, the sedimentary histories are corrected for the effect of compaction during the geological past (Fig. 3, solid lines) (DU ROUCHET, 1980; STEGENA et al., 1981; FALVEY & DEIGHTON, 1982).

- Based on present borehole temperatures, the geotherms for each 10° C round interval are constructed in the time-depth section (Fig. 4, left) with the present heat flow during the geological past. The constancy of heat flow during the past does not result in parallel and equidistant straight lines; it is possible to take into consideration the probable changes with time and depth of thermal conductivity of the layers, with the aid of the lithology and burial history of the borehole.

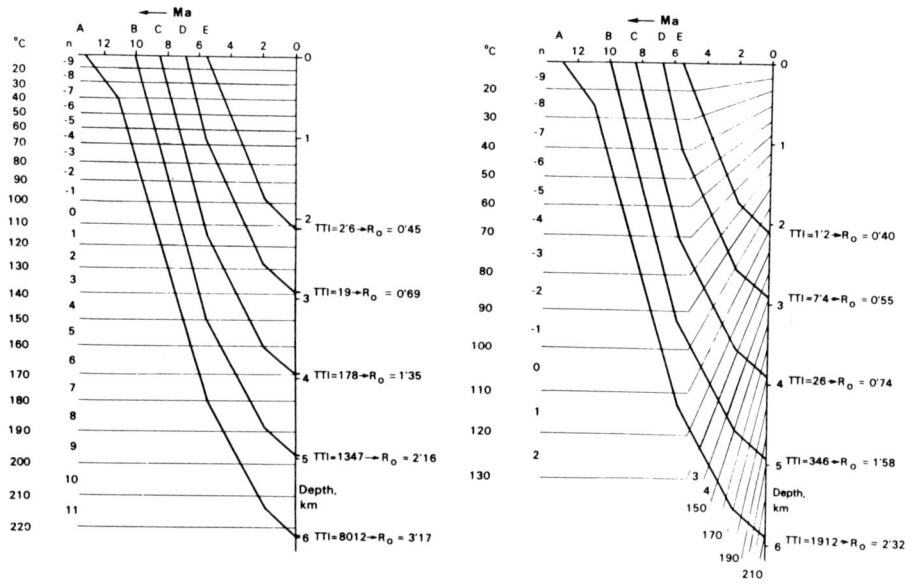

Fig. 4. Calculated TTI values for the borehole HOD assuming that the heat flow density was constant through the sedimentary history (left), and that the borehole was heated up during the last 5 Ma for the present heat flow value (STEGENA et al., 1981)

- After calculating the TTI values for each layer of the borehole the TTI-s are transformed to R_o values (Fig. 4), using the correlation of WAPLES (1980) (Fig. 2).

- These calculated R_o values are compared with the R_o values measured in the borehole. The discrepancy between calculated and measured values is attributed to the variations of heat flow during the geological past. Using plausible hypotheses, one makes a change in the past heat flow (Fig. 4, right) and repeats the comparison of R_o values calculated from TTI-s and measured R_o values, till a good fit between calculated and measured vitrinite reflectances is achieved (Fig. 5).

- Fig. 5 shows two boreholes of the Pannonian basin with heat flow histories calculated independently. Both boreholes gave the same result: the measured vitrinite reflectances are compatible with the assumption that the Pannonian basin has had a low heat flow (~ 50 mW/m²) before 5 Ma, and 5 Ma ago the heat flow began to increase (linearly?) to its present value (~ 100 mW/m²).

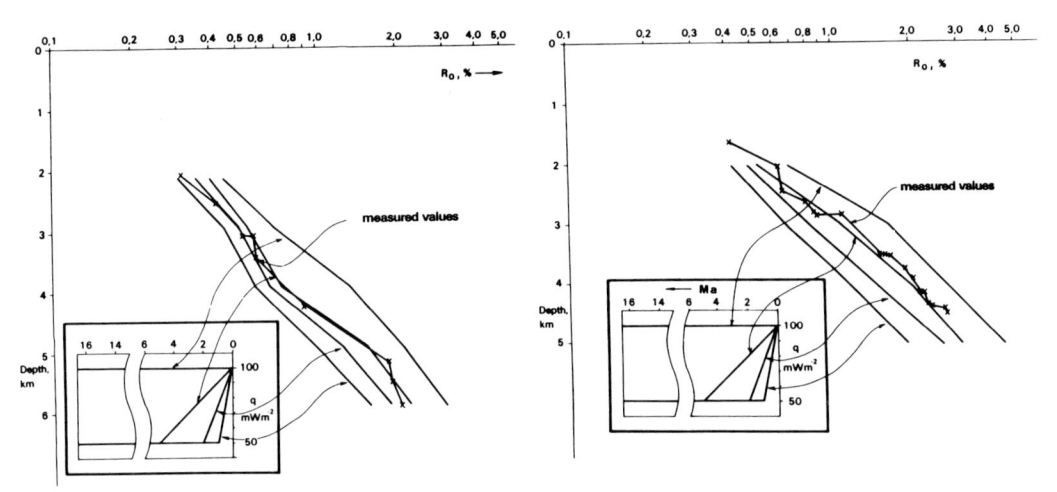

Fig. 5. The measured vitrinite reflectances in the borehole HOD and DER (both in the Pannonian basin) and the vitrinite reflectances calculated from the following heat flow stories: the heating-up of the boreholes began at ∞, 5, 2, 1 Ma ago (after STEGENA et al., 1981).

The above scheme serves better to understand the principles of the paleogeothermal calculations, but does not present a final solution of the question. There are some fundamental problems in the oil geochemistry which are not solved satisfactorily and which can influence the above sketched model.

- It became usual to assume that increases in vitrinite reflectance values were valid indicators of the extent to which organic matter maturated and oil generation had occurred (WAPLES, 1983). However, there is an uncertainty in some R_o measurements, because the values have a wide spread, and sometimes it is hard to distinguish low reflecting resinite and high reflecting fusunite from vitrinites (HO, 1978). During the beginning of oil generation, bitumen impregnations lower the vitrinite reflectance. In all red-coloured rocks organic matter is oxidized; in limestones vitrinite is very rarely preserved and if it occurs, the reflectance value differs from the value of vitrinite in the same rank. RONSARD & OBERLIN (1984) suggest that, as with any other electronic property of any solid, reflectance depends on three parameters: chemical composition, atomic structure and microstructure. The same value for reflectance can thus be measured for materials different in their microstructure and chemical composition, which can be of different ranks or not. They suggest the use of transmission electron microscopy (TEM) by using successive heat treatment in an inert atmosphere to 1000° C, which better characterizes the maturation of organic materials.

- It is generally supposed that pressure does not have a significant effect on the maturation of organic matter and on the amount of hydrocarbon generated. It is to be noted however, that the role of pressure in oil generation has never been examined properly (WAPLES, 1983).

- The maturation of organic matter exhibits a very complex process, involving a lot of parallel chemical reactions with various activation energies, and the whole process can hardly be described by a first-order kinetic expression (SIEVER, 1983). This was also shown by pyrolysis experiments (CUMMINGS & ROBINSON, 1972). LASAGA (1981) has compiled a table of activation energies for geochemical reactions that shows a range from less than 4 kJ/mole to more than 400 kJ/mole. TISSOT (1969), TISSOT & ESPITALIÉ (1975), TISSOT et al. (1975), and JÜNTGEN & KLEIN (1975) have modelled the thermal alteration of kerogen with a set of first-order rate equations,

$$\frac{d\,n_{ki}}{dt} = -n_{ki}\,a_i\,\exp(-\frac{E_i}{RT}) \qquad i = 1,2,\ldots 6$$

where n_{ki} is the mass function, a_i is the frequency factor, E_i is the activation energy of the i-th kerogen. If it is integrated over the thermal history of any horizon, the generated petroleum and the maturity of organic matter can be calculated. This process although giving a better theoretical approximation, is hardly applicable for paleogeothermal applications.

LOPATIN (1971) tested his model on a very difficult well, Münsterland 1/FRG. Recalibration of Lopatin's method with larger and more reliable data sets (WAPLES, 1980; KETTEL, 1981) has verified the general validity of the model itself, but has modified Lopatin's original TTI-vitrinite reflectivity correlation. LOPATIN & BOSTICK (1973) and LOPATIN (1976) later suggested some improvements to the original scheme. LOPATIN (1976) used fewer and larger temperature intervals; instead of $\Delta T = 10°$ C,

$$\Delta T = \frac{1.37 T^2}{E - 1.37 T}$$

(T in Kelvin, E activation energy = 42 kJ/mole).

This formula gives 15° C for ΔT at $T < 80°$ C, 20° C at 80° $C < T < 120°$ C, 25° C at 120° $C < T < 170°$ C, and 30° C for ΔT above 170° C.

The diagenesis of organic matter accelerates exponentially with temperature. In the whole process, the time which the layer under consideration passed away at maximum temperatures, plays a decisive role. HOOD et al. (1975) worked out a model, in which the period spent within 15° C of the rock's maximum paleotemperature was taken into consideration. For the maturation of organic matter, and indirectly, for the vitrinite reflectivity, they created a scale of thermal maturity called the "level of organic metamorphism" (LOM), which is controlled only by the maximum temperature survived by the layer, and by the "effective heating time" spent by the layer within 15° C of the rock's maximum temperature (Fig. 6). Staplin's similar scale (TAI, thermal alteration index) is based on microscopic structure variation and the colouring of organic debris.

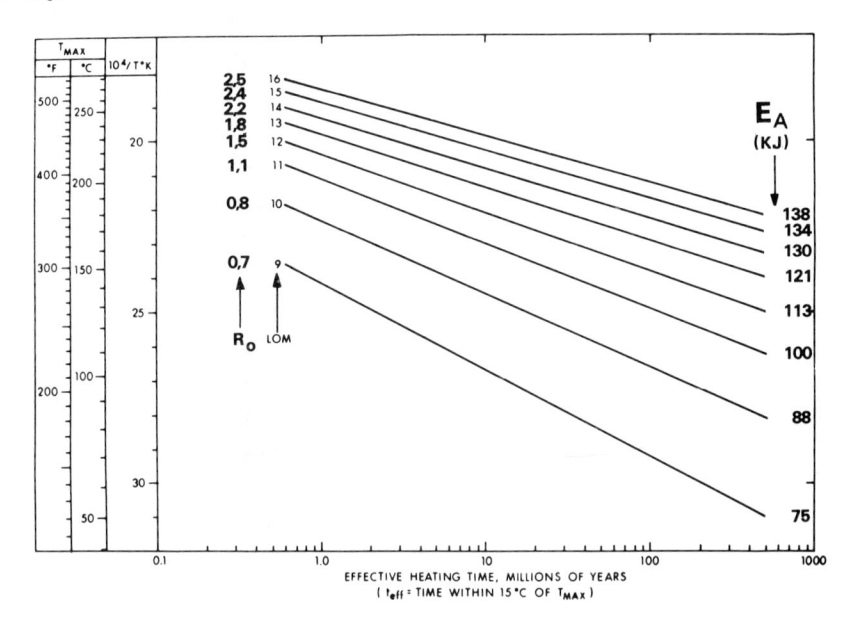

Fig. 6. Relation of LOM and R_o to maximum temperature and effective heating time (after HOOD et al., 1975, modified)

PUSEY (1973) suggested that maximum paleotemperatures can be obtained accurately from ESR (electron spin resonance) analysis of kerogen. The ESR is sensitive to free radicals; the number of free radicals increases as kerogen is subjected to increasing temperatures, and kerogen free radicals are stable through geologic time. The ESR geothermometer was calibrated by obtaining data from cores of Tertiary basins believed to be actively subsiding and so satisfying the highly probable assumption that samples from these basins are now at maximum temperature since burial. But ESR signals are not only dependent on temperature but are also subject to variations in kerogen type, diagenetic changes in kerogen, weathering and geologic time.

PRICE (1982) improves the idea that vitrinite reflectivity depends first of all on maximum temperature. A plot of R_o versus present temperature from a number of areas that have not undergone major geologic mutilation, increases in a strictly linear fashion (r= 0.97) yet burial times for these different areas range from 0.3 to 240 Ma. He suggests that some geochemical postulates are in error and that time has little effect on organic maturation. It appears that vitrinite reflectivity can be used as an absolute paleogeothermometer from 20° C to at least 400° C.

MIDDLETON & FALVEY (1983) propose, for simplicity, that maturation (C) and R_o are related by the equation

$$\ln R_o = A + BC.$$

Empirical studies give A = -2.275 and B = 0.177.

For maturation C, they accept Lopatin's original idea with insignificant modification (ΔT = 10.2 instead of 10° C) and for simplicity use the logarithm of the previously given integral

$$C = \ln \int_0^t 2^{T(t)/10.2} dt$$

(as used by ROYDEN et al., 1980 and DE BRAEMAEKER, 1983).

Equations combine to give an equation relating R_o to temperature as a function of time:

$$(R_o)^a = b \int_0^t \exp[c\, T(t)]\, dt$$

where a = 5.635, b = 2.7·10^{-6} Ma^{-1} and c = 0.068° C^{-1}.

Given the thermal history of an organic sediment T(t), this equation can be used to predict the vitrinite reflectivity of the sediment after a time t. Nor does WELTE & YÜKLER's (1981) equation add more to that formulated by LOPATIN (1971) and WAPLES (1980):

$$R_o\,[\%] = 1.301\, \lg(TTI) - 0.5282.$$

BUNTEBARTH (1978) tried to calculate paleogeothermal gradients, as far as possible without theory. It is clear that a relationship exists between the coal rank, measured by the mean optical reflectivity of vitrinite (Rm), and the integral of depth and duration of burial. A correlation has been evaluated between the square of vitrinite reflectivity and the burial history:

$$Rm^2 \sim \int_0^{t_1} z(t)\,dt$$

(z depth, t time, t_1 means that the calculation can be restricted to a part of the whole burial history).

Furthermore, it is clear that, in this relationship, the coal rank is proportional to a function of the geothermal gradient,

$$Rm^2 = 1.16 \cdot 10^{-3} \exp(0.068\, dT/dz) \int_0^{t_1} z(t)\,dt.$$

Fig. 7 shows measured Rm values in some boreholes in the F.R.G. as a function of burial history ($\int_0^{t_1} z(t)\,dt$).

Geothermal gradients measured at present in the four boreholes in Fig. 7 allowed the calibration of the empirical equation. The applicability of this equation for other areas is investigated in BUNTEBARTH & MIDDLETON (this volume).

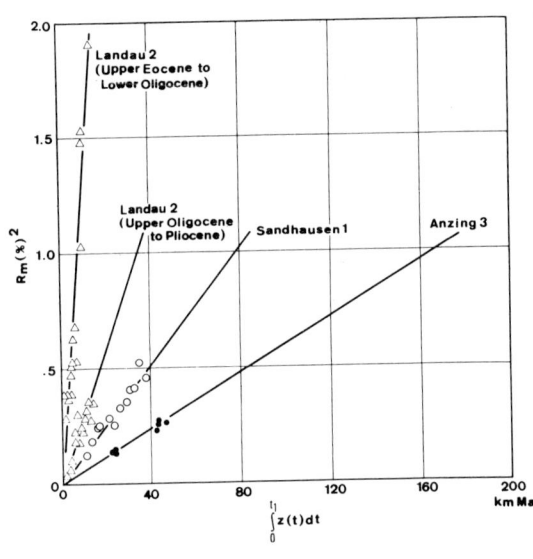

Fig. 7. Relation between the mean vitrinite reflectance (Rm) and the integral of depth and time, in four boreholes of the Upper Rhinegraben (BUNTEBARTH, 1979)

Some case histories:

Based on maximum measured vitrinite reflectivity data (HACQUEBARD, 1977) and burial history of 28 wells lying in the Central Prairies Basin, Canada, MAJOROWICZ & JESSOP (1981) estimated a lower average paleogeothermal gradient (27 mKm^{-1}) for the early Oligocene time than the present day one (30.6 mKm^{-1} in average) (Fig. 8). For the calculation, they used the method of KARWEIL (1956) with BOSTICK's (1973) modifications and the method proposed by HOOD et al. (1975).

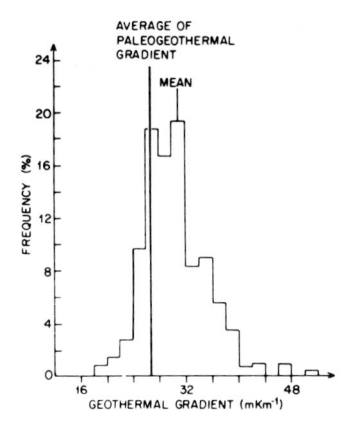

Fig. 8. Average of the paleogeothermal gradients with the histogram of present geothermal gradients, in the Central Prairies basin, Canada (MAJOROWICZ & JESSOP, 1981)

EGGEN (1984) worked with a lot of vitrinite reflectivity data but present heat flow estimations only. He stated that in the Viking Graben (Norwegian North Sea) the calculated paleo heat flow density (approx. 55 mWm^{-2}) fits well with the present heat flow estimation (50 - 60 mWm^{-2}); on the flank of the Viking Graben, however, an average paleo heat flow close to 50 mWm^{-2} has to be assumed in order to obtain the observed maturity, while the present day estimation lies at 70 mWm^{-2}.

WANG JI-AN et al. (this volume) found that the coalification gradient increases from 0.25 to 0.65 % R_o/km from middle to early Eocene, in the western part of Liaohe oil field, North China, and a coalification gradient of about 0.4 was determined in the early Tertiary sediments of the Central Hebei oil field. KARWEIL's (1956) and LOPATIN's (1971) methods were used for paleotemperature reconstructions. In contrast, RYBACH (1984) gives 0.09 - 0.05 % R_o/km coalification gradients for the Northern Alpine Foreland (Molasse basin).

ROYDEN & KEEN (1980) predict R_o values for the sediments of the Nova Scotia and Labrador shelves, based on theoretically derived thermal evolution, and on LOPATIN's theory. A similar work was carried out by ROYDEN et al. (1980), for the Falkland Plateau and for three places of the North Atlantic.

BUNTEBARTH (1983,1985) estimated the paleotemperature gradient as well as the heat flow density in a few sedimentary basins in the F.R.G.

In the Ruhr Basin, the foredeep of the Rhenish Variscan mountains, for which many data are available (BUNTEBARTH et al., 1982), the heat flow decreased during Westphalian C from about 125 mW/m² to about 105 mW/m². Because of the low thermal conductivity of the coal, the temperature gradients reached mean values of 79° C/km before, and 65° C/km after this decrease in heat flow data obtained.

The thermal regime of the back-deep of the Rhenish Variscan mountains, that of the Saar Basin, is nearly the same as that of the Ruhr Basin during the Westphalian (BUNTEBARTH, 1983). A similar high heat flow is indicated in other European Carboniferous basins, e.g. ROBERT (1985).

Within the Lower Saxony Basin, the Upper Carboniferous coal beds were heated by intrusive bodies during the Upper Cretaceous - about 200 Ma after sedimentation. Model calculations that take the cooling of the Massif of Bramsche into account, indicate that temperature gradients between 60 and 80° C/km existed within the coal bearing strata. From model calculations, prior to the magmatic heating, the temperature gradient did not exceed 30 to 40° C/km during maximum burial (BUNTEBARTH, 1985).

The paleogradient derived for the borehole Urach 3 (Swabian Alb) fortuitously agrees with the measured present day gradient. The paleogradient of 43° C/km corresponds to Cretaceous to Lower Tertiary times, because coalification ended prior to the Upper Tertiary uplift (BUNTEBARTH & TEICHMÜLLER, 1982).

The thermal regime of the middle Upper Rhine Graben changed during the Tertiary. Temperature gradients during the Lower Tertiary were higher than those during the Upper Tertiary. The values ranged, respectively, from 48 to 78° C/km, and from 34 to 50° C/km (BUNTEBARTH, 1978). A direct relationship may exist between volcanic activity in the Graben, and the high thermal gradients, both documented for the period immediately after the opening of the Graben.

From these few data, it can be tentatively concluded that the Upper Cretaceous was a time of widespread high thermal gradients. Furthermore, high gradients also existed during the Upper Carboniferous in northern Germany, and during the Lower Tertiary in southern Germany.

In contrast to the problems of vitrinite reflectivity enumerated, these results seem to be realistic.

To avoid the difficulties with vitrinite reflectivity, McKENZIE (1981) proposed a new idea:

Some of the problems relating to the empirical relations suggested by LOPATIN (1971), WAPLES (1980) and others could be avoided if chemical reactions involving only one molecular type which occur during the maturation of the organic material, were to be identified. MACKENZIE & McKENZIE (1983) have investigated the rates of three reactions which occur before and during the early stages of oil formation. Two of the reactions are isomerization reactions, at C-20 in a sterane and at C-22 in a hopane hydrocarbon; the third reaction converts C-ring monoaromatic to triaromatic steroid hydrocarbons. All three reactions were assumed to be first order and monomolecular; the isomerization reactions are reversible, with a rate of conversion of the R to the S form of 1.174 and 1.564 resp., while the aromatization reaction was assumed to be irreversible.

This method excels by its clear theoretical (thermodynamic) principles, the Arrhenius equation is certainly valid for these reactions. The problem, however, is that frequency factor and activation energy cannot be determined in the laboratory or only very inaccurately, because of the slowness of the reactions. Because of this, McKENZIE's (1978) stretching theory for the evolution of sedimentary basins was used to calibrate the reactions. This theory involves a thermal history, which can be derived sufficiently accurately from the burial history. Based on chemical analyses of North Sea and Pannonian Basin cores, and using more or less determined or hypothetical stretching models, the kinematics of the three reactions were determined in Table 1.

Table 1. Rate parameters of three reactions (MACKENZIE & McKENZIE, 1983)

	Frequency factor (s^{-1})	Activation energy ($kJ\,mol^{-1}$)
Isomerization of steranes	$6 \cdot 10^{-3}$	91
Isomerization of hopanes	$16 \cdot 10^{-3}$	91
Aromatization of steroid HC-s	$18 \cdot 10^{14}$	200

Fig. 9 shows the results for the Pannonian basin, which are in a certain agreement with the theoretical curves derived with the assumption that the stretching rate, β, is 2 (SCLATER et al., 1980). A similar study was carried out by HOFFMANN et al. (1984) for the Malakam Delta, Kalimantan, Indonesia, and SAJGÓ et al. (1983) for the Pannonian basin. Articles of SAJGÓ & LEFLER (this volume) give detailed information about the applicability of some marker reactions to paleogeothermal determinations.

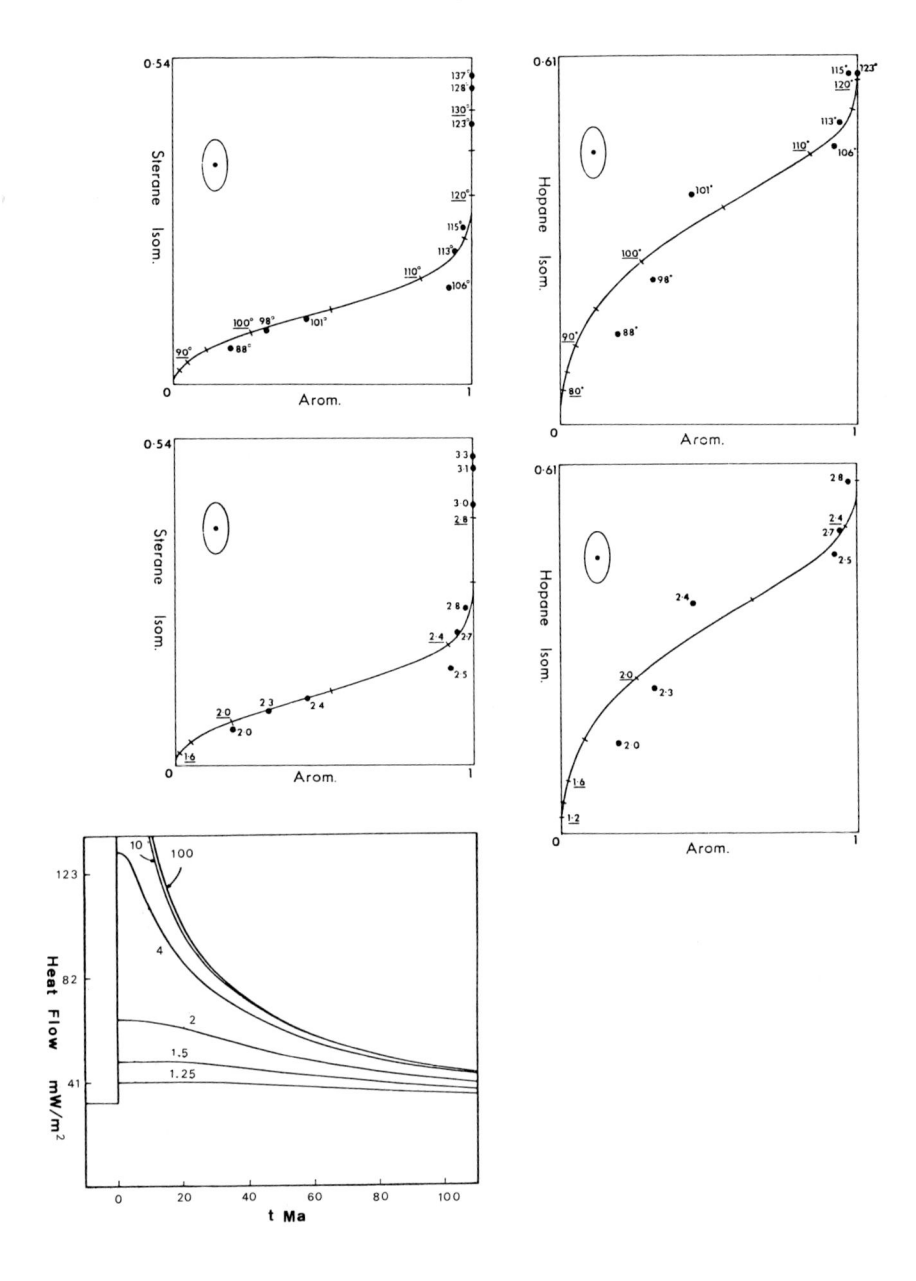

Fig. 9. The extent of sterane and of hopane isomerization as a function of steroid hydrocarbon aromatization for samples of the borehole HOD in the Pannonian basin. The curves are calculated theoretically, based on the thermal history of the basin from McKENZIE's (1978) stretching theory. The basin is assumed to have been formed by sudden extension (ß = 2), 15 Ma ago. The marks on the curves are present temperatures at 5° C intervals (above) and present depths at 200 m intervals (below) (MACKENZIE & McKENZIE, 1983). The left lowest diagram shows the approximate thermal history which belongs to various ß values.

2. Fluid inclusion thermometry

In nearly all minerals, whether ores, rock forming minerals or others, small amounts of fluids are entrapped in the host crystal which preserve the physical and chemical conditions of the surrounding medium during the time of the crystal growth. It is generally assumed that no subsequent change in the entrapped material takes place (LEMMLEIN, 1956; ROEDDER, 1967).

At the time of entrappment, the fluid inclusion is a homogeneous phase constituting mainly of water, salt (in general sodium chloride) and some amount of carbon dioxide, and also silicate melt. Since the thermal expansion of the fluid/melt is greater than that of the mineral, a vapour bubble is formed within the cavity when the temperature decreases.

The fluid inclusions are formed by crystal growth when the advancing faces, edges and corners of the growing crystal are disturbed (primary inclusions), by fracturing and healing of crystals during mechanical disturbances or by overgrowth of a crystal (secondary inclusions) (PAGEL & POTY, 1983).

From rock or mineral samples, thin sections (~80µm) are prepared which are mounted on a glass plate and polished on both sides. The thin section is heated up under a microscope by using a heating stage (e.g. OHMOTO & RYE, 1970). At a certain temperature, the bubble disappears in the inclusion. The heating is then reversed to cooling until the entrapped fluid/melt becomes an inhomogeneous phase again. The temperature at this condition is measured and called the homogenization temperature. This particular temperature is related to the temperature of formation. However, the pressure of formation has to be involved. Generally, an increase in pressure requires a greater temperature to complete homogenization (ROEDDER, 1967; SIGURDSON, 1974; POTTER, 1977; BOWERS & HELGESON, 1983a,b). The pressure correction is different for silicate melt inclusions with shrinkage vapour bubbles (only ~20° C/kb) and for more compressible fluids with water and carbon dioxide (ROEDDER, 1982).

Since the fluid inclusions contain a substantial concentration of sodium chloride, the pressure correction can be applied, if this concentration is known. The thermodynamic properties of aqueous solutions are affected remarkably by the concentration of sodium chloride. The pressure correction reported by POTTER (1977) is based on volumetric properties of the $NaCl-H_2O$ system. The correction by BOWERS & HELGESON (1983a,b) works at pressures above 50 MPa and high temperatures from 350 to 600° C; additional to the graphs of the ternary system, FORTRAN programs are given to generate the pressure correction (BOWERS & HELGESON, 1985) which is based on a modified REDLICH & KWONG (1949) equation of state matching the pressure-volume-temperature data reported by GEHRIG (1980).

Prior to the pressure correction, the sodium chloride content must be analyzed. The salinity of the fluid in inclusions can be estimated by the depression of the freezing point during cooling. The higher the salinity, the higher is the depression of the freezing point. ROEDDER (1962) reports the freezing point data for pure solutions of sodium chloride. Applying this data to fluid inclusions, the salinity of the brine can be given in sodium chloride equivalent only, since the fluid consists of other constituents, and this mixed salt solution causes some uncertainty in the determination of the sodium chloride content (ROEDDER, 1976). The freezing point of the inclusion is determined during cooling of the sample using cooled nitrogen gas. When the inclusion is frozen, the gas flow is reduced so that the crystals begin to melt. The melting temperature of the last ice crystal determines the freezing point of the inclusion. The pressure is taken either as the actual pressure corresponding to the burial depth of the sample, or as the paleo-pressure which is estimated from the sedimentary history.

CURRIE & NWACHUKWU (1974) and MAGARA (1978) used this principle for determination of paleo-geothermal gradients in Canadian Cardium sandstone as follows: Thin sections were made from fracture-filling materials (mainly quartz) of sandstone cores from 5 boreholes of a single reservoir. Those quartz fillings that contained fluid inclusions were heated and microscopically observed. The ranges of homogenization temperatures and calculated paleogeothermal and measured present geothermal gradients are shown in Table 2.

Table 2. Ranges of homogenization temperatures and calculated geothermal gradients of Cardium sandstones (by CURRIE & NWACHUKWU, 1974 and MAGARA, 1978).

Well	Homogenization temperature (°C)	Maximum burial depth geothermal gradient (mK/m)	Near present gradient (mK/m)	Present gradient (mK/m)
A	45 - 108	38	33	
B	46 - 100	35	25	
C	50 - 85	33	33	32
E	51 - 84	36	33	
F	51 - 88	33	31	

It was supposed that the highest homogenization temperature of a core refers to the quartz filling formed at maximum temperature and the lowest homogenization temperature originates from quartz fillings formed at lower ("near present") temperatures.

The paleo-pressures were determined by sedimentary history, taking the rock compaction (using sonic log) into consideration. Fig. 10 shows the diagram of interpretation: the measured maximum homogenization temperatures (A-F) are followed along the line of specific volume of water till the calculated paleo-pressures are reached; this gives the paleotemperatures A' - F'. For the lowest homogenization temperatures (a - f), the present pressure of cores (calculated by depth and density) were applied.

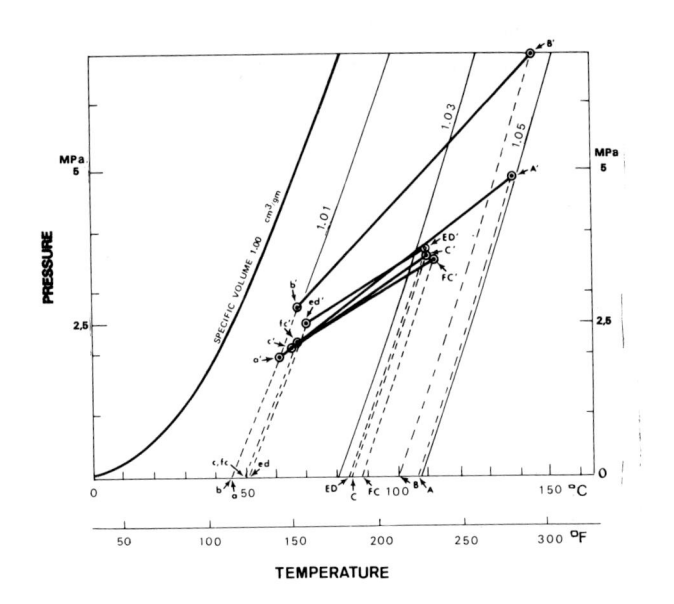

Fig. 10. Graphs showing ranges of homogenization temperature of Cardium sandstone in five (A - F) boreholes in western Canada and the interpreted paleotemperatures (a' - f', A' - F') (from MAGARA, 1978, modified)

The homogenization temperature method excels in its simplicity, and has been applied since the last century. However, there were some uncertainties which biased the results, and which may have been the reason why the method was seldom applied till the fifties when the technique and the foundations were improved.

Inclusions can contain not only water but many other materials: mixed salt, oil, hydrocarbons, carbon dioxide, etc. (FREY et al., 1980; POTY & PAGEL, 1983). FREY et al. (1980) demonstrate that the fluid composition changes continuously in fissure quartz of the external parts of the central Alps, from higher hydrocarbon, through methane to water-bearing fluid inclusions. The change of the fluid composition is dependent mainly on the temperature increase during the progressive Alpine metamorphism. The continuous change of composition may cause some systematic uncertainties in both the pressure correction (ZAGORUCHENKO & ZHURAVLEV, 1970) and the estimation of the formation temperature. ROEDDER (1962,1963) proposed to begin the heating experiment at low (-35° C) or even very low (-180° C) temperatures (POTY & PAGEL, 1983), because the freezing temperatures are useful for discriminating among gas, liquid and supercritical fluid, and among liquid water, liquid oil and liquid carbon dioxide.

Fluid inclusions may also re-equilibrate during burial, or could possibly leak, but experimental studies of quartz show that most fluid inclusions will not decrepitate at an internal overpressure lower than 800 bars and that the smallest can withstand overpressures as high as 4 kilobars (TUGARINOV & NAUMOV, 1970; LEROY, 1979).

In many cases, fluid inclusion studies give minimal temperatures and only occasioally actual temperatures of inclusion formation. If the fluid was homogeneous at the time of trapping, the homogenization temperature will be a minimum temperature. The trapping temperature of the fluid in basins which are 5 - 7 km deep may be higher by a value up to 80° C, in comparison to the fluid inclusion temperature.

POTY & PAGEL (1983) suggest that fluid inclusion techniques seem to give more detailed data than organic matter and illite crystallinity techniques.

Case histories

PAGEL (1975), using the fluid-inclusion microthermometry of detrital quartz grains of sandstones, received a 35° C/km geothermal gradient for sedimentary basin in Athabasca (Canada) "which is characteristic for active sedimentary basins" (Fig. 11). The fluids of inclusion exhibited a NaCl concentration of 30 %.

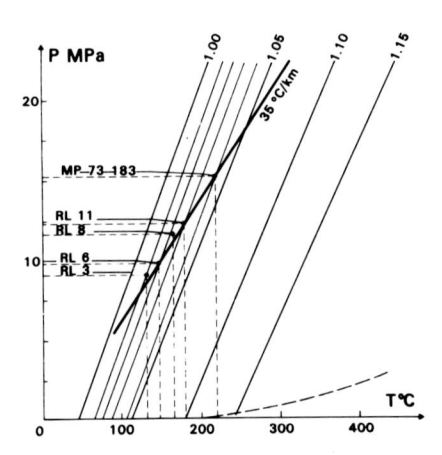

Fig. 11. PVT diagram for a 30 weight % NaCl solution and the calculated paleotemperatures by microthermometry of detrito quartz grains from Athabasca sandstones of five (RL3 etc.) Canadian boreholes (after PAGEL, 1975)

VISSER (1982) studied petroleum source rock from Venezuela by fluid inclusion thermometry. Thin (< 80 µm) sections were made, polished on both sides and mounted on coverslips with epoxy resin to give the necessary support. The heating-freezing experiments were performed on a stage as described by POTY et al. (1976). The fluid inclusion data of secondary inclusions in quartz- and calcite-filled veins showed that the maximum diagenetic temperature is in good agreement with the actual measured formation temperature of 157° C (Fig. 12).

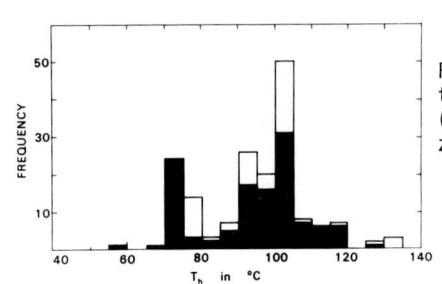

Fig. 12. Frequency diagram of homogenization temperatures in calcite (black) and quartz (white), in a petroleum source rock from Venezuela (VISSER, 1982)

Based on the primary inclusions - which were entrapped during the initial crystal growth - in diagenetic quartz overgrowths, TILLMAN & BARNES (1983) stated that the host rock (sandstone) temperatures in the Northern Appalachian Basin vary from 176 to 147° C, with an average of 155° C; the burial depth, at the end of the Paleozoic or early in the Mesozoic, were at depths of 3.5 km (= 40° C/km). The average geothermal gradient measured today in central and western New York is about 25° C/km. The milky and clear calcite samples belong to the Oswego fault system and mirror a wide range of temperature of later hydrothermal events (Fig. 13).

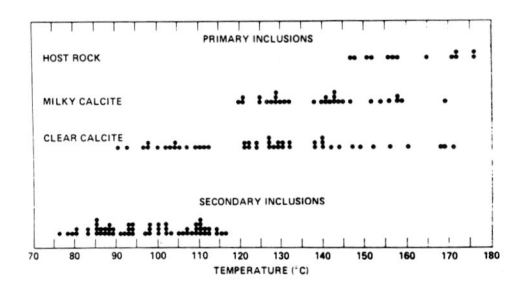

Fig. 13. Fluid inclusion temperature distribution in Northern Appalachian Basin sandstones (TILLMAN & BARNES, 1983)

Fluid inclusion studies in active geothermal fields are reported, e.g., by BROWNE (1973) and BROWNE et al. (1974) from the Broadlands/New Zealand; by FRECKMAN (1978) from the Salton Sea geothermal field, Imperial Valley/USA; and by TAGUCHI et al. (1979) from Hatchobaru/Japan.

In the Broadlands geothermal area it was found that the borehole temperatures coincide with the homogenization temperature in quartz which indicates a constant thermal regime. The homogenization temperatures in quartz, anhydrite, and calcite of the Salton Sea area are closely associated with the borehole temperatures, but vein calcite exhibits some fluctuations which may be due to episodic changes in faulting, fracturing and subsequent fluid flow at different temperatures. The homogenization temperatures in quartz and anhydrite in the Hatchobaru geothermal field indicate a higher thermal regime than was measured, which indicates that the climax of heat flow existed in the past, and the geothermal reservoir is already cooling down.

3. Geochemical thermometry

In a saturated porous rock a solution equilibrium is established after a certain time and at a given temperature. An equilibrium can exist either within the pore fluid among different isotopes, or between the rock matrix and the pore filling. The temperature dependence of the reactions considered here is used to estimate the temperature of the environment where the equilibrium is established.

The geochemical thermometry is mainly applied to fluid and vapour dominated geothermal systems and is based on the following requirements:

(a) freely available elements (species) in the (rock-water) system
(b) equilibrated reactions in the deep reservoirs
(c) slow re-equilibration during upward migration.

The fractionation of stable isotopes of various elements, e.g. in solution, has been recognized as a temperature indicator because of the temperature dependence of the fractionation factor (e.g. HOEFS, 1973; O'NEIL, 1979). The isotope ratios of elements hydrogen (D/H) and oxygen ($^{18}O/^{16}O$) are used in different reactions for temperature estimation.

The analyzed isotope ratio reflects the equilibrium condition of an isotope exchange reaction, as for example

$HD + H_2O \rightleftharpoons H_2 + HDO$ (HULSTON, 1976; ARNASON, 1976; GIGGENBACH & LYON, 1977)
$CH_3D + H_2 \rightleftharpoons CH_4 + HD$ (HULSTON, 1976; ELLIS & MAHON, 1977)
$H_2^{16}O + (HS^{18}O^{16}O_3)^- \rightleftharpoons H_2^{18}O + (HS^{16}O_4)^-$ (HULSTON, 1976; ELLIS & MAHON, 1977)

The time needed for re-equilibration is only a matter of days or weeks, respectively, or months for the latter reaction. Because of the shortness of this time, these isotope exchange reactions are unsuitable for paleothermometry.

As well as isotope exchange, the solution equilibrium between the rock matrix and the pore fluid is used for temperature determination. Only a very small part of the rock component enters into solution. Individual minerals and amorphous components such as glass phases in eruptive rocks or opal have very different solubility. Rock-forming minerals such as quartz and feldspar are much less soluble than salts.

The SiO_2 content of thermal water is frequently used to determine the temperature prevailing in subsurface water reservoirs. Even though there are many limiting factors for the validity of equilibrium conditions, the SiO_2-thermometer yields a useful temperature estimation. However, temperatures calculated from this thermometer are generally too low (KOLESAR & DEGRAFF, 1978). Inaccuracies arise, e.g., because of varying solubilities of different SiO_2 polymorphs such as quartz, chalcedony, cristobalite and amorphous SiO_2 (FOURNIER & ROWE, 1966; FOURNIER, 1981). Experience shows that low temperatures (to $T \approx 100°$ C) should be calculated with the chalcedony solubility and higher temperatures with that of quartz.

Even though the SiO_2 thermometer has been used for some time with success to estimate an actual reservoir temperature (FOURNIER & ROWE, 1966; MAHON, 1966; ARNORSSON, 1975; ELLIS & MAHON, 1977), it has considerable disadvantages which are mostly related to the absolute content of dissolved SiO_2. These disadvantages do not favor this method for paleotemperature estimations, although the re-equilibrium of an existing silica-water equilibrium is a slow process.

The advantage of the Na-K and the Na-K-Ca thermometer is that instead of the absolute content, the ratios of their concentrations are used to estimate the equilibrium temperature (WHITE, 1965; FOURNIER & TRUESDELL, 1978; PACES, 1975; ELLIS & MAHON, 1977, FOURNIER, 1981). The temperature obtained from the Na, K and Ca content of groundwater is often higher than that which is determined by the SiO_2 content. SiO_2 can precipitate out during the mixing of cold water with thermal water. If the thermal source is within a sedimentary basin, the salinity of the prefilling fluid is important. Also considerable perturbation of the thermometer can be caused by monomineralic salt deposits.

Although the re-equilibration is very slow, slower than for silica, the SiO_2, Na-K and Na-K-Ca thermometer are of little use for paleothermometry. The problem arises because water, being mobile, migrates up- and downward through layers, and the dissolved silica or Na/K reflects first the temperature-differences caused by the migration process; these geothermometers are therefore used for the determination of the "basic-temperature". In very special cases only, if water migration is excluded or exactly known, silica and K/Na geothermometers can be used for paleogeothermometry.

KHARAKA et al. (1980) have reported ^{18}O isotopic analyses from subsurface waters from a well of Brazoria County, Texas. The strata under consideration were deposited in marine or near-marine settings. During burial, formation water typically becomes enriched in ^{18}O as a result of reactions with the surrounding sediment. A plot of oxygen isotope values for calcium carbonate with depth shows a trend toward lighter values at greater depth (MILLIKEN et al., 1981), supporting the water isotope data, but with a big scatter. ELDERS et al. (1984)'s results from the ^{18}O content of calcites from sandstones in Cerro Prieto geothermal field boreholes show similar scatter but are useful (Fig. 14). The four typical curves of Fig. 14 reflect at first the cold and hot water migration processes and give an important contribution to deciphering the past fluid migration processes and hence, to estimating the paleogeotemperatures.

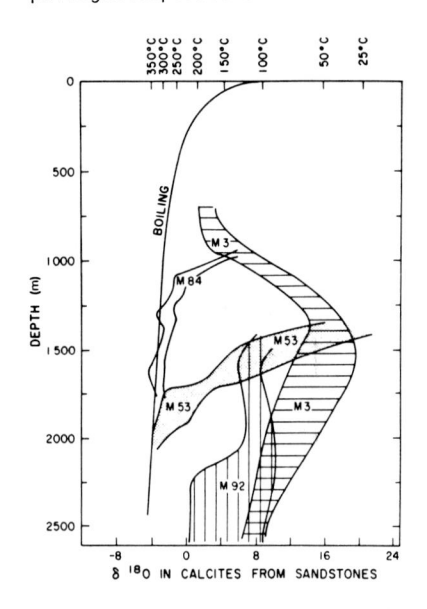

Fig. 14. Ranges of $\delta^{18}O$ in calcites from sandstone samples recovered from four different wells characteristic of the Cerro Prieto geothermal field. The shaded areas show the range of values measured. The "boiling" curve shows the $\delta^{18}O$ of calcite in equilibrium with boiling water with $\delta^{18}O = -8.33\ ‰$ (ELDERS et al., 1984).

Fractionation of stable isotopes of the elements hydrogen, carbon, oxygen, and sulfur in two minerals are also used for temperature determination (HOEFS, 1985). This isotope thermometry has become well established since the classic paper of UREY (1947). The principle of this thermometry is that the partitioning of two stable isotopes of an element between two minerals depends on the temperature. The isotope ratio of compounds (1) and (2) is compared with that of a standardized sample and the relative difference is known as "delta" δ-value. The fractionation factor α is the ratio of the isotopic composition of two compounds (1) and (2) and is approximately the difference of their δ-values:

$$10^3 \ln \alpha_{(1,2)} \approx \delta_{(1)} - \delta_{(2)}$$

Since the fractionation factor α is temperature dependent (BOTTINGA & JAVOY, 1973):

$$\alpha \sim \exp(x_1/T^{x_2})$$

where $x_{1,2}$ are constants and T is the absolute temperature, the difference of the δ-values is a function of T^{-x_2}. Experimental results are consistent with $x_2 = 2$ for high temperature (O'NEIL, 1979), so that

$$100 \ln \alpha = A/T^2$$

The constant A must be known in order to determine the formation temperature of two coexisting minerals.

The temperature determined is taken as the last equilibrium temperature. However, an equilibrium temperature cannot be assumed from all samples at all terraines. The isotopic fractionation can change due to chemical alteration or recrystallisation of minerals and other kinetic effects which inhibit a complete re-equilibration (HOERNES & HOFFER, 1985). Isotopic temperatures may sometimes indicate a thermal condition or an event which is difficult to interpret, e.g. when isotope exchange reactions took place during retrograde metamorphism (HOEFS, 1985). The application of additional paleothermal methods may confirm or assist in understanding uncertain results in this case. Generally, data from the Salton Sea geothermal field/California (FRECKMAN, 1978) and from mines (HOEFS, 1985) demonstrate that fluid inclusion analysis and isotopic geothermometry are in good agreement.

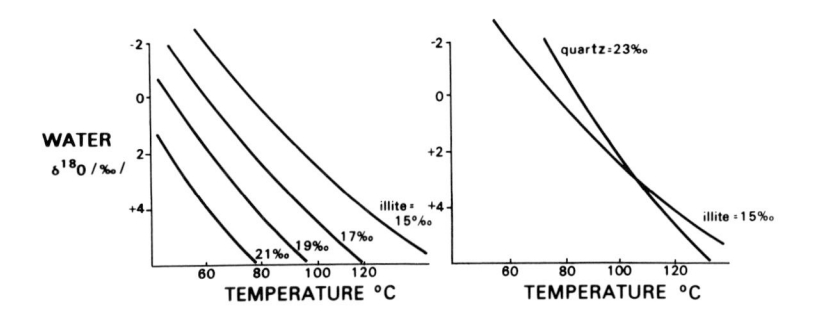

Fig. 15. Relationship between water and mineral $\delta^{18}O$ values (SAVIN & LEE, 1984). Left: lines indicating range of possible conditions for formation of illites with $\delta^{18}O$ values of +15, +17, +19 and +21 per mille, at various temperatures, and $\delta^{18}O$ values of the ambient water. Right: $\delta^{18}O$ values of two cogenetic minerals uniquely define the temperature of formation and the isotopic composition of ambient water. The cogenetic mineral pair have been formed at about 106° C in the presence of ambient water with $\delta^{18}O$ of about 3 per mille.

SAVIN & LEE (1984) argue that most minerals once formed in oxygen isotopic equilibrium with the ambient water, are extremely resistant to subsequent isotopic exchange with environmental waters at sedimentary and most diagenetic temperatures, except when they undergo chemical or mineralogical alteration. As a result, the $^{18}O/^{16}O$ ratios of minerals can provide information about their conditions of formation. If the $^{18}O/^{16}O$ ratios of two cogenetic mineral phases (e.g. illite and quartz, Fig. 15) can be measured, then both the temperature of formation and the $^{18}O/^{16}O$ ratio of the ambient water can be calculated.

4. Transformation of minerals in sedimentary rocks

In sedimentary rocks, some authigenic minerals undergo a diagenesis during burial within the uppermost few kilometers of depth. The low temperature of up to 200° C causes an alteration of clay minerals, silica polymorphs, and zeolites. The transformation temperature of each series of authigenic minerals in diagenetically altered argillaceous sediments can be used for evaluating the geothermal history (AOYAGI, 1979).

Most of the investigations are based on the transformation of clay minerals (BURST, 1969; PERRY & HOWER, 1970,1972; AOYAGI et al., 1975; HOWER et al., 1976). Montmorillonite as an expanding clay mineral transforms at increasing temperature and pressure due to water and CO_2 loss, to potassium-poor smectite/illite which are not expandable. The interlayer water of montmorillonite which is released during the transformation to illite is considered to play an important role in petroleum migration (POWERS, 1967; BURST, 1969; PERRY & HOWER, 1972). AOYAGI & ASAKAWA (1977) argued that both interlayer and interstitial water expelled during the diagenesis were responsible for oil migration.

The processes resulting in the diagenesis of montmorillonite begin at the depth corresponding to 80° C geotemperature and generally end at 120° C (BURST, 1969; JONES, 1970). A part of montmorillonite remains below this depth because the absorbable potassium available is not enough for the montmorillonite-illite transformation (JONES, 1970), or transforms in a phase which is to be grouped with the pyrophyllites (WEISS & ROLOFF, 1965).

AOYAGI & ASAKAWA (1984) report a temperature of 104° C for the transformation of montmorillonite to mixed-layer montmorillonite (smectite)/illite, and one of 137° C for mixed layer minerals to illite which can be assumed for the Neogene argillaceous sediments of Japan.

In the middle Upper Rhine Graben/FRG, the transition zone between Montmorillonite and mixed layer minerals is generally found in the Graue Schichtenfolge Formation (Oligocene) in a depth corresponding with a temperature of about 70° C. The depth ranges between 700 m and 1300 m. HELING & TEICHMÜLLER (1974) excluded from that a significant influence of overburden pressure.

Another measure of the diagenesis of clay minerals is the "illite crystallinity" (KÜBLER, 1967) which gives a continuous scale for the degree of diagenesis. Its value is the width at half amplitude of the first-order illite basal reflexion measured with an X-ray diffractometer. The application to the very lowgrade metamorphism in external parts of the Central Alps show an obvious relation, however being divergent at different sites between illite crystallinity, coal rank and fluid inclusion data (FREY et al., 1980). The investigations seem to exhibit a substantial influence of local conditions in each area. It has also been thought that the montmorillonite-illite transformation may actually be a kinetically controlled process (WAPLES, 1980), and the progress in transformation reflects the thermal history of the sediment. WAPLES (1980) stated a certain correlation between the temperature-time-index (TTI) as calculated for diagenesis of organic matter and the proportion of expandable clay layers (Fig. 16). Samples lying significantly left of the line (in a thermally immature region), probably represent material which contained less than 100 % expandable layers when it was originally deposited. This is in agreement with the generally accepted view, that the montmorillonite/illite ratio also depends on the paleogeographic environment. It is believed that primary illite is associated with regressive, and montmorillonite with transgressive phases (CHAPMAN, 1973).

ELDERS et al. (1984) determined the progressive zones of hydrothermal alteration minerals in sandstones at the Cerro Prieto geothermal system, Baja California, Mexico, based on a lot of deep boreholes. The wide temperature-bands and their overlap give some idea of the suitability of this method for paleothermal applications (Fig. 17); for the Cerro Prieto field however, where very high temperature changes occurred as a consequence of a young (~50000 y) hot (~1000° C) shallow (~6 km) thermal plume, ELDERS et al. (1984) constructed realistic thermal histories, for the last 50000 years.

A transformational sequence of zeolites is reported by AOYAGI & KAZAMA (1977) for paleotemperature determinations. It has been recognized in Neogene argillaceous rocks from deep boreholes in Japan. This alkali zeolite reaction series comprises 4 zones:

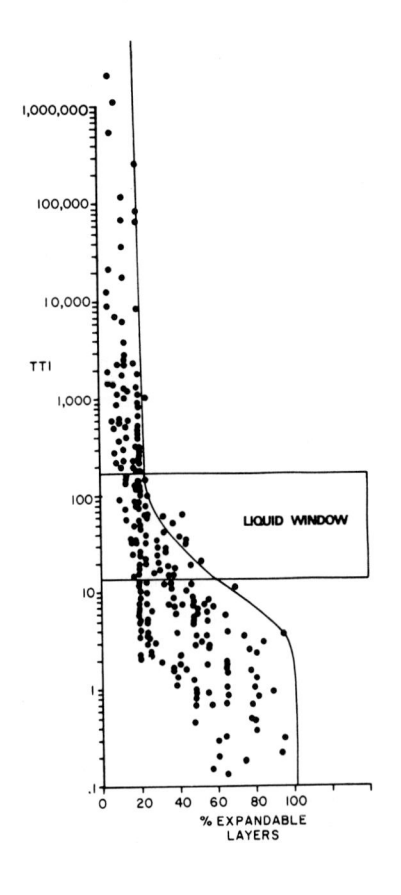

Fig. 16. Time Temperature Index of maturity versus % expandable layers in mixed-layer clays (WAPLES, 1980)

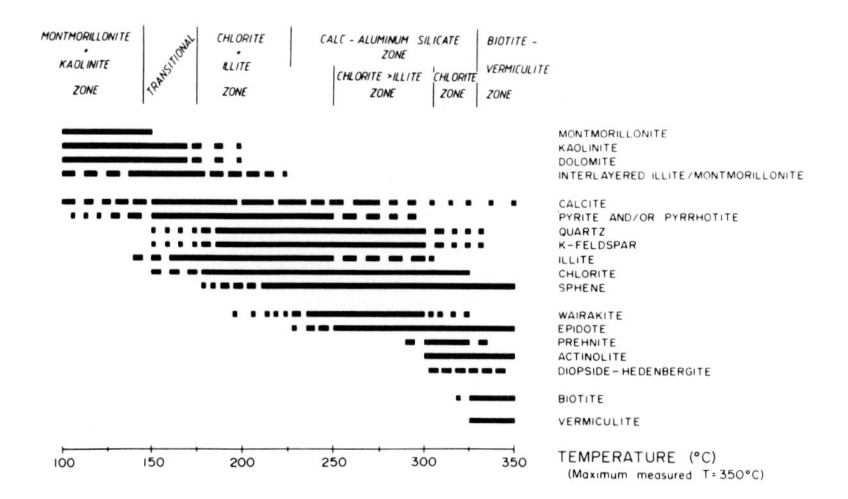

Fig. 17. Temperature ranges of zones of hydrothermal alteration minerals in the sandstones of Cerro Prieto geothermal field (ELDERS et al., 1984)

1) silicic (volcanic) glass
2) clinoptilolite ± mordenite + cristobalite
3) analcime + quartz
4) albite + quartz

which are supplemented with a calcic zeolite series yielding heulandite in zone 3 and laumontite in zone 4.

IIJIMA et al. (1984) report a transformation temperature to the next zone of 53° C, 85° C, and 122° C, respectively, which has been estimated from the bottom hole temperature of the Miti-Kuromatsunai borehole in Hokkaido, Japan. A paleothermal gradient of 31° C/km has been evaluated, whereas the present gradient has a value of 52° C/km. The area was uplifted and a thickness of 600 m was eroded since the last 0.5 Ma.

The transformation is essentially time-dependent which has been deduced not only from field observations but also from experimental and theoretical studies (AOYAGI & ASAKAWA, 1984).

For evaluating paleotemperatures, the silica minerals give a transformational sequence during diagenesis of sedimentary rocks. At lower temperatures amorphous silica is stable which is transformed to low-temperature cristobalite and finally to low-temperature quartz with increasing temperature (MITSUI & TAGUCHI, 1977). The transformation temperatures are 45° C and 56° C, respectively.

All three transformation series of minerals (Fig. 18) are used to evaluate the paleotemperature gradient of the Miti Hamayuchi borehole in northern Hokkaido, Japan (AOYAGI & ASAKAWA, 1984). The depth of the first appearance of authigenic minerals is: 650 m low-temperature cristobalite; 950 m clinoptilolite; 1500 m low-temperature

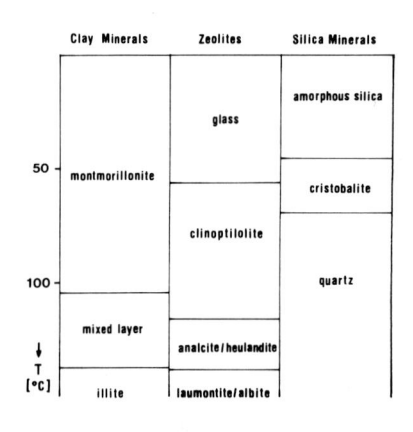

Fig. 18. Transformation temperature of clay minerals, zeolites, and silica polymorphs in Neogene argillaceous sediments of Japan (after AOYAGI & ASAKAWA, 1984)

quartz; 2800 m mixed layer minerals. The transformation temperature of each boundary is 45° C, 56° C, 69° C, and 104° C, respectively. From these data, a paleotemperature gradient of 27° C/km is calculated which was valid during the Neogene.

Another 16 deep boreholes in the Niigata basin, Honshu, Japan were analyzed by AOYAGI & ASAKAWA (1984). The paleogeothermal gradient during the Neogene ranged from 19 to 36° C/km. The paleogeothermal gradient in oil and gas fields of the area generally cover the upper range between 30 and 40° C/km.

5. Conodont Color Alteration

During the Paleozoic era a group of animals lived in the sea, whose remains are to be found in great numbers in many sedimentary rocks of that era. Because these hard, mineral remains resemble teeth, they are called conodonts. The nature of the animal group is, however, unknown. They have no descendant in their evolution, since they became extinct during the Upper Cretaceous, and any essential parts of the animal which could have found its place in the classification of the animal kingdom such as the soft, perhaps tissuelike parts cannot be reconstructed. Although this group of animals, of which several hundred forms are distinguishable, is rather enigmatic, their remains, i.e. the conodonts,are of great importance in geology for dating and mapping the sedimentary layers in which they occur (LINDSTRØM, 1964). The size of the conodonts rarely exceeds 1 mm, lying mainly between 0.1 and 1 mm.

Conodonts were widely spread in the seas in which they occurred since Late Cambrium. They passed through an evolution which produced so many characteristic and widespread forms that they are valuable for stratigraphic mapping. The most prosperous period in the evolution culminates perhaps in the Late Devonian. During Carboniferous and Permian, the conodonts occurred less frequently. A relative climax can be seen from Middle Triassic sedimentary rocks. In most parts of the world the conodonts became extinct before the Cretaceous age. This group of animals outlasted a period of more than 300 Ma.

The conodont remains of the animals consist of calcium phosphate with some minor amounts of carbonate, fluor, and sodium (PIETZNER et al., 1968). As this carbonate apatite is rather resistent to physical and chemical changes of the environment, the conodonts are well preserved, even into the garnet-grade metamorphic facies where they underwent temperatures as high as 500° C (EPSTEIN et al., 1977).

Conodonts are abundant in black shales as well as in limestones. The more fine-grained the rocks, the better is the chance to find enough material. LINDSTRØM (1959) suggested that the conodont frequency is inversely proportional to the sedimentation rate. There might be some environmental influence which favours conditions of quite, perhaps warmer sea water than that of streaming water in which sand and other coarse-grained material are deposited. Under favourable conditions several thousand conodonts can be found per kilogram rock material. In coarse-grained material, e.g. sandstone, or in limestone deposited at a high rate, conodonts are absent or as rare as a few examples per kilogram.

In paleogeothermics, the color of the conodonts is of special interest. It ranges from pale yellow, through different brown tones, to black. The coloration seems to be due to the carbonization of some trace amounts of organic matter, probably amino acids as reported by PIETZNER et al. (1968). This carbonization is essentially dependent on the temperature and on the time at which the conodonts were exposed to that effective temperature.

EPSTEIN et al. (1977) reported heating experiments with conodonts, and introduced a color alteration index in which five steps in color alteration can be discriminated by comparing the conodont under the microscope with a color standard. The laboratory heating of pale yellow conodonts of the color alteration index "1", i.e. the first step with which the alteration starts, comprises a temperature range from 300° C to 600° C and a duration of the heating up to 50 days. The experiments allowed EPSTEIN et al. (1977) to draw an Arrhenius-plot for each conodont alteration index (Fig. 19), from which the effective temperature of natural colored conodonts can be read.

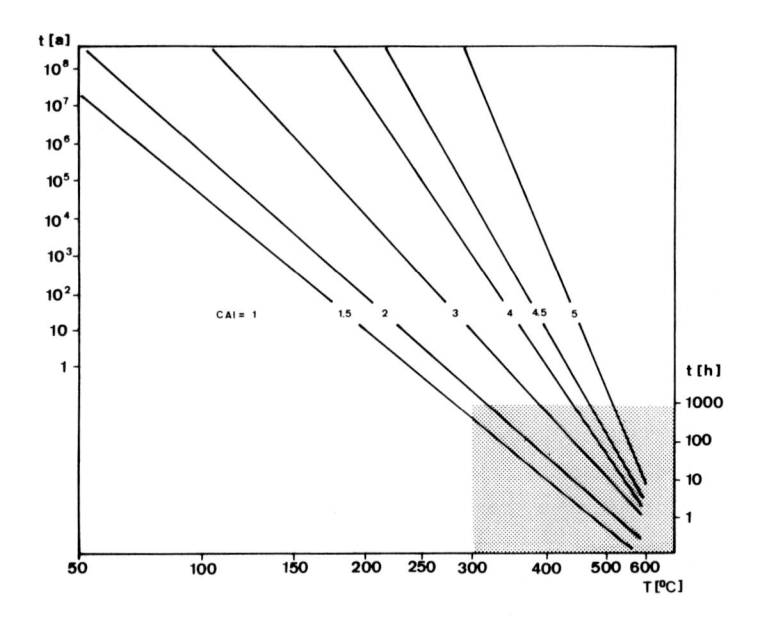

Fig. 19. Arrhenius-plot of the conodont color alteration using the color alteration index (CAI) as a parameter. The shaded area shows the field of experimental control (after EPSTEIN et al., 1977).

EPSTEIN et al. (1977) reported that conodont colors correspond with the colors from field collections, and that their alteration is progressive, cumulative, and irreversible. The temperature and time dependence has been mentioned above. In principle, the Arrhenius equation is supposed to describe the process which causes the color alteration.

Since the density of the color also depends on the thickness of the specimen which is observed, there is a limited accuracy in this method. From this, EPSTEIN et al. (1977) concluded that temperature intervals below 50° C cannot be discriminated using the qualitative color alteration index.

The validity of the Arrhenius equation seems to be much more founded in this case than it is for coaly particles. The organic compounds, which might be amino acids only, are less complex than those derived from plant remains. Furthermore, the environment within the conodonts, i.e. within the carbonate apatite, is rather constant, whereas a great variety exists for plant remains in different sedimentary rocks. Other advantages of the conodont color alteration are that the method can be applied for Cambrian rocks, in which conodonts are already abundant, but vitrinite is still rare, and that the temperature range is wider and reaches 600° C instead of about 350 - 400° C for coaly matter.

The conodont color alteration can also be used to support vitrinite reflectance methods. From using the first method, the thermal history of a basin can be estimated using marine carbonate rocks, whereas the second method estimates the thermal history from more clastic rocks. Both methods seem unaffected by tectonic events, neither folding nor faulting.

Case history

EPSTEIN et al. (1977) applied the conodont color alteration to Middle Ordovician rocks from Monterey, Virginia in the Valley and Ridge province. They found a color alteration index of 4 to 4.5. According to geologic observations, the maximum time for heating was 270 Ma which corresponds with a temperature of 185 - 220° C applying Fig. 19.

The earliest possible time for uplift in this area could be during Late Pennsylvanian which results in a heating time of up to 210 Ma. The corresponding temperature ranges from 190 to 230° C, which is not much different from the first result and which demonstrates that the time dependence is less important for long heating periods. In the vicinity of Monterey, it is supposed that a 4770 m thick sequence of sedimentary rocks covered the Middle Ordovician. If a surface temperature of

20° C is taken into account, a thermal gradient between 43 and 52° C/km is estimated for the time of the deepest burial, i.e. before Late Pennsylvanian.

6. Radiometric dating

Radiometric age determinations are based on the assumption that radioactive systems such as K-Ar, Rb-Sr and U spontaneous fission are closed during a time span which is then determined as being the age. This means that neither the diffusion of daughter nuclides nor the population of fission tracks changes. The age which is determined from each system denotes the age since closure. As different systems close and open at different temperatures, combined radiometric age determination techniques can be applied to evaluate the thermal history of a rock sample. The time for age determination begins when the rock passes through a certain temperature during cooling. Several temperatures and ages are determined from a rock sample, so that the cooling history can be evaluated.

The closure temperature in biotite has been estimated as $300 \pm 50°$ C for the K-Ar and the Rb-Sr system (PURDY & JÄGER, 1976). HAMMERSCHMIDT & WAGNER (1983) ascertained a value of $330 \pm 20°$ C for the K-Ar system by diffusion experiments. The closure temperature in muscovite is somewhat higher (about 350° C) (WAGNER et al., 1979).

Fission tracks in apatite and zircon are annealed at lower temperatures. HAMMERSCHMIDT et al. (1984) concluded from experimental data that in geological ages temperatures between 50 and 150° C are sufficient for the annealing of fission tracks in apatite. According to the time-dependence, the annealing temperature of the Odenwald crystalline basement/FRG is taken as 100° C and that of the Central Alps as 125° C (WAGNER, 1968; WAGNER & REIMER, 1972). Within the geologically young Cerro Prieto geothermal field, Mexico, SANFORD & ELDERS (1981) assume, corresponding with a duration of heating of 1000 to 10,000 years, 170° C as the temperature annealing of fission tracks in detrital apatite.

With the combined application of both age determination techniques, the apatite fission track and the Rb-Sr system in biotite yields in the Urach III borehole, southern Germany, a paleotemperature gradient of 55 to 60° C/km which was active during the Cretaceous period (HAMMERSCHMIDT et al., 1984), whereas the present temperature gradient has a value of 40° C/km (HAENEL & ZOTH, 1982).

ZAUN & WAGNER (1984) determined the annealing temperature of fission tracks in zircon using the paleotemperature data of the Urach III borehole. It was found that the fission tracks are stable up to a temperature of 130° C. The closing temperature, that is the temperature of which half of the fission tracks anneal, is determined as $170 \pm 20°$ C.

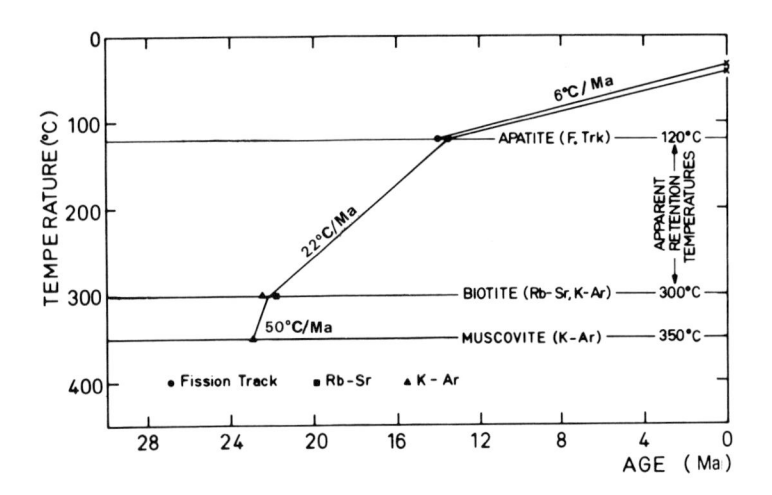

Fig. 20. Cooling history of the Bergell intrusive, Central Alps, based on the retention temperature of Rb-Sr and K-Ar biotite, K-Ar muscovite and apatite fission track systems (after WAGNER et al., 1979, modified)

WAGNER et al. (1979) combined various dating methods to reconstruct the cooling history of the Bergell intrusive, Central Alps. An U-Pb age from zircons of the intrusion of 30 Ma is reported. After which, the muscovite K-Ar system passed through the closing temperature of 350° C about 23 Ma ago, and the biotite K-Ar as well as Rb-Sr system passed through the retention temperature of 300° C about 22 Ma ago. The apatite fission track method gives an age of 14 Ma. From these data, the ages vs. temperatures give the reconstructed cooling rates as shown in Fig. 20.

TEMPERATURE HISTORY OF THE EARTH'S SURFACE IN RELATION TO HEAT FLOW

N.J. SHACKLETON
University of Cambridge, Godwin Laboratory for Quaternary Research
Free School Lane, Cambridge CB2 3RS England

In order to interpret the temperature gradient below the earth's surface in terms of heat flow, the simplest assumption is that the concept of a long-term mean temperature is valid and that this mean temperature may be measured at a depth below the surface sufficient that seasonal and year-to-year variations are negligible. However, this assumption is not appropriate when temperature profiles in deep sections are analysed. The purpose of this contribution is to point to some of the sources of information for cases when the past surface temperature record must be taken into account.

It is on the glacial-interglacial timescale that we have the most detailed information regarding the scale and tempo of temperature variation at the earth's surface. As regards the scale, perhaps the most useful information concerns the last glacial maximum around 18 ka (thousand years) ago. CLIMAP (1976,1981) mapped the distribution of surface temperatures over the oceans. The more recent publication (CLIMAP, 1981) analyses the distribution of the glacial temperature anomaly in some detail. In itself, information regarding temperature at the sea surface is not useful in relation to heat flow studies. However, a modelling study (GATES, 1976) showed that the relatively modest temperature anomalies reconstructed for the oceans did in fact give rise to significantly larger temperature anomalies on the continents that are consistent with other geological evidence. This in turn means that even in areas for which geological data are lacking, an atmospheric modelling study is likely to provide a reliable estimate of the temperature anomaly associated with glacial maximum conditions. This is rather reassuring in view of the vast and disparate nature of the literature on Pleistocene palaeo-environments.

The tempo of glacial-interglacial temperature changes is known to be controlled by changes in the earth's orbital geometry (MILANKOVICH, 1941; HAYS et al., 1976). Thus to a first approximation the pattern of past temperature changes is better modelled in terms of these astronomical changes rather than being regarded as a square-wave alternation of glacial and interglacial states. IMBRIE (1985) indicates a simple means of achieving such a reconstruction. Current work is aimed towards a better understanding of the geographical distribution of the climatic response to the variations in the three relevant astronomical variables ("tilt", period about 40 ka; "precession", period about 22 ka, and "ellipticity", period about 100 ka).

Most Pleistocene sections from continental regions are very discontinuous and hence only yield temporally isolated temperature estimates, rather than continuous temperature curves suitable for integrating. One method that is being used in order to obtain time-averaged temperature anomalies is to make use of the temperature dependence of the racemization rate of aminoacids (MILLER et al., 1983), and a project is under way to evaluate the temperature anomaly over the past glacial cycle over Europe. This type of data may prove more valuable for heat-flow workers than spot estimates of temperature based on fauna, flora or morphological features such as ice-wedge casts, which may represent extreme climates.

Particular care must be taken in temperature estimation in some areas. For example, very cold conditions with mean annual temperatures 10 - 20° C cooler than today prevailed close to the ice sheet margins (van der HAMMEN et al., 1967) due largely to the cold katabatic winds flowing off the ice sheets. On the other hand beneath the ice sheet the temperature may have been much higher and extensive areas were probably at pressure melting point (although some areas would have been frozen to the bedrock). Ice sheet modelling studies are being directed towards an evaluation of the temperature conditions under the major ice sheets.

Arid areas may also provide unforeseen complications resulting from changes in precipitation. For example the very extensive lake areas in now-arid areas (STREET & GROVE, 1976,1979) must have significantly changed temperature distributions, while the change from forest to savannah or desert (HAMILTON, 1976; TRICART, 1975) which affects shade cover, would also have a major impact on temperature at ground level.

Until recently it was believed that the temperature at the sea floor in the deep oceans did not change significantly on a glacial-interglacial scale. However, recent work has required a re-evaluation of this assumption and it now seems likely that even in the deep Pacific, bottom water was cooler by about 1.5° C in glacial times than today, with a larger anomaly in the Atlantic (DUPLESSY et al., 1985; SHACKLETON & CHAPPELL, 1985). The simplest interpretation of the data now available suggests that colder conditions in the deep sea occupied up to 90 percent of late Pleistocene time, so that for the purposes of heat-flow modelling it might be appropriate to consider that deep temperatures were colder for the past million years up until a step-warming about 10 ka ago.

There are also areas in the deep sea, where special care must be taken in estimating past temperatures. A particular case is the Mediterranean Sea. Today deep water in this enclosed basin forms in winter in the area near the French coast. In glacial times the winter sea-surface temperature was much colder in this area so that Mediterranean deep water was also much colder. No detailed evaluation of the benthic stable isotope data has been undertaken, but it is likely that glacial deep-water temperatures in the Mediterranean were at least 5° lower than today.

On a shorter timescale, temperature anomalies associated with events such as the "little ice age" of the 17th to 19th centuries are investigated by a combination of historical and geological methods. Although the anomalies are smaller than those associated with glacial times, they should not necessarily be ignored; significant mountain glacier advances are recorded (DENTON & KARLEN, 1977) in some areas and mean annual temperature anomalies of the order 1° C are not uncommon. A good source for changes on this timescale is the review by LAMB (1977).

It is widely believed that in the early Holocene about 6 ka ago temperatures were higher than today globally. A recent detailed review (WEBB, 1985) of the data on which this belief is based suggests that it is probably not the case that global average temperatures were higher at that time, although there were probably temperature maxima of regional extent.

On the timescale long compared with glacial-interglacial changes, even more dramatic changes have taken place. Over the past 50 Ma high latitudes have cooled by more than 10° C, as has the deep ocean. At the same time it appears that low latitudes were actually somewhat cooler 50 Ma ago than today (SAVIN, 1977; SHACKLETON, 1985). The past 40 Ma have been anomalous by comparison with the previous tens of million years by up to about 10° C.

ISOTOPE GEOTHERMOMETERS

J. HOEFS
Geochemisches Institut der Universität Göttingen
Goldschmidtstr. 1, D-3400 Göttingen, F.R. of Germany

Abstract

The best available isotope thermometers of the elements oxygen, sulfur, carbon and hydrogen are summarized, and some applications to metamorphic rocks, to sulfide ore deposits and to geothermal systems are discussed.

Whether or not isotope geothermometers are generally applicable depends whether or not isotope equilibrium is established, and if the existence of isotope equilibrium may be recognized. An inherent danger is the tendency to regard calculated temperatures as estimates of peak thermal conditions. However, the temperatures determined represent the last isotope equilibration, below which no further isotope exchange takes place. This temperature often coincides with fluid loss from the geological system.

Only those mineral pairs can be used as geothermometers where temperature calibrations exist. From the three different approaches a) theoretical calculation, b) calibration on an empirical basis, c) experimental determination, the latter seems to be the most promising, although considerable disagreement exists between some published calibration curves. However, with new sophisticated techniques on hand (CLAYTON et al., 1983a,1983b) this difficulty may be overcome in the near future.

Introduction

Isotopic thermometry has become well established since the classic paper of Harold UREY (1947) on the thermodynamic properties of isotopic substances. The partitioning of two stable isotopes of an element between the mineral phases can be viewed as a special case of element partitioning between two minerals. There are, however, quantitative differences between these two exchange reactions, the most important being that isotope partitioning is more or less pressure independent, which represents the greatest advantage relative to the numerous other geothermometers.

Recently, RUMBLE (1982), however, argues that changing pressure has a significant influence on isotopic fractionations in rocks. The pressure effect arises because changing pressure causes changes in the proportions of volatile species in fluids, which in turn leads to changes in fractionation between bulk fluid and bulk rock.

The necessary condition to apply the different geothermometers is isotope equilibrium. Conclusions concerning the nature and the extent of isotope equilibrium are influenced by the criteria used to test for attainment of equilibrium and the spatial scale over which measurements have been made.

In a mineral assemblage of n-phases we can obtain n-1 independent temperatures, one temperature for each mineral pair. If each mineral pair gives concordant temperatures, we can be nearly certain that isotope equilibrium was attained and that equilibrium was frozen in at the same temperature in every mineral. A disadvantage of the concordant temperature method is that it tacitly assumes temperature calibrations are accurate.

Isotopic compositions of two compounds A and B are expressed as δ-values. The fractionation factor α of an isotope exchange reaction is related to the measured δ-values through the following approximation:

$$\delta_A - \delta_B = \Delta_{A-B} \approx 10^3 \ln \alpha_{A-B}$$

Considering the exchange of only one atom, the fractionation factor α is equivalent to the equilibrium constant K. Theoretical studies show that the fractionation factor α, for isotope exchange between minerals, is a linear function of $1/T^2$, where T is in degrees Kelvin, at crustal temperatures. BOTTINGA & JAVOY (1973) were the first to show that isotopic fractionations between mineral pairs can be expressed in terms of the equation

$$1000 \ln \alpha = A/T^2$$

which means that for a temperature determination factor A has to be known.

Temperature calibrations

Three different methods have been used to determine the equilibrium constants for isotope exchange reactions:

a) calculation from statistical mechanical theory, which is especially suitable for gas reactions

b) experimental determination in the laboratory

c) calibration on an empirical basis.

The latter method is based on the idea that the calculated "formation temperature" of a rock in which other minerals are also present, serves as a calibration to the measured fractionations of other minerals, providing that all minerals were at equilibrium. However, since there is evidence that totally equilibrated systems are not very common in nature, this empirical calibration should be abandoned.

The theoretical calculation of isotope fractionation factors is exceedingly difficult, because all vibrational frequencies of the crystalline lattice must be taken into account. Therefore the most promising approach seems to be the experimental determination of isotope fractionation factors.

In principle, the experimental determination of isotope exchange equilibrium constants can be carried out simply by holding the phases at a fixed temperature. By a suitable choice of isotopic compositions of the starting minerals, it is possible to approach equilibrium from opposite directions, thus satisfying the classical criterion for equilibrium.

However, the driving forces for the exchange reactions are small, and rates of exchange are often very low. In such cases, a variety of techniques has been used to facilitate exchange, summarized by CLAYTON (1981):

1) recrystallization of a very finely ground powder
2) crystallization of a gel or glass
3) crystallization as a result of polymorphic phase transition
4) synthesis of a new phase by cation exchange
5) complete mineral synthesis.

All of these techniques depart from an ideal exchange experiment in that there are driving forces for reaction other than the differences in isotopic composition. These obvious limitations result in various calibration curves for which significant discrepancies exist.

Such experimental determinations have been most extensively carried out for oxygen isotopes. Recently, the development, at the University of Chicago, of high-pressure direct-exchange techniques using a three isotope approach, has significantly extended the range of systems which can be studied, and has provided a coherent set of mineral pair fractionations (MATSUHISA et al., 1978,1979; MATTHEWS et al., 1983a,1983b) (Table 1).

The three oxygen isotope method has initial mineral-water $^{18}O/^{16}O$ fractionations that are close to equilibrium, but initial $^{17}O/^{16}O$ ratios that are removed from equilibrium. The extent to which a system has approached isotopic equilibrium is accurately determined from the changes in the $^{17}O/^{16}O$ mineral-water ratios, and the corresponding near to equilibrium $^{18}O/^{16}O$ ratios are then extrapolated to determine equilibrium.

Table 1. Coefficients A for silicate-pair fractionations (after MATTHEWS et al., 1983a, 1983b)

	Ab	Cc	Jd	Zo	An	Di	Wo	Mt
Qz	0.5	0.5	1.09	1.56	1.59	2.08	2.20	6.11
Ab	-	0.0	0.59	1.06	1.09	1.58	1.70	5.61
Cc	-	-	0.59	1.06	1.09	1.58	1.70	5.61
Jd	-	-	-	0.47	0.50	0.99	1.11	5.02
Zo	-	-	-	-	0.03	0.52	0.64	4.55
An	-	-	-	-	-	0.49	0.61	4.52
Di	-	-	-	-	-	-	0.12	4.03
Wo	-	-	-	-	-	-	-	3.91

$$(1000 \ln \alpha_{A-B} = \frac{A}{T^2} 10^6)$$

Abbreviations: Qz - quartz, Ab - albite, Cc - calcite, Jd - jadeite, Zo - zoisite, An - anorthite, Di - diopside, Wo - wollastonite, Mt - magnetite

Oxygen

There is a debate about the extent of isotope equilibrium in igneous and metamorphic rocks. A survey of the literature data on oxygen isotope fractionations by DEINES (1977) lead him to conclude that only a small portion of the data truely indicates isotope equilibration of the whole assemblage. On the other hand, HOERNES & FRIEDRICHSEN (1978), besides others, argued that many metamorphic rocks attained isotope equilibrium during metamorphism. The most common examples of disequilibrium assemblages are obviously due to retrograde isotope exchange. When a fluid phase is present during the cooling period of a hot rock, it will effect continuing isotope exchange between coexisting minerals down to temperatures well below the maximum temperatures. Isotope exchange will be terminated by the effective cessation of diffusion in the particular phase. Isotope exchange might cease at different temperatures according to differing diffusion parameters of the respective minerals. Inspection of the published data suggests the following general pattern

$T_{quartz-magnetite} > T_{quartz-muscovite} \approx T_{quartz-biotite} > T_{quartz-feldspar}$
due to the different diffusion rates of the different minerals (FREER & DENNIS, 1982).

With respect to retrograde isotope exchange two different models can be imagined, which cannot be resolved by measuring the isotopic composition of a particular component (Fig. 1). Isotope temperatures in the upper model report the last isotope exchange between a mineral and a fluid phase during the retrograde cooling of a thermal event M1. Alternatively, in the lower model temperatures are reset during a subsequent lower grade thermal event M2. In both cases, the resetting most probably occurred during recrystallization of the minerals, but this is not a necessary condition to apply this model.

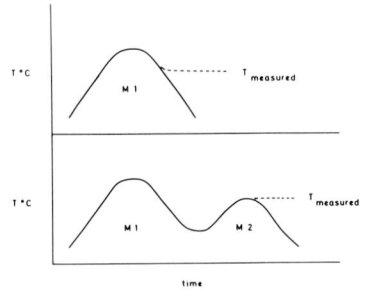

Fig. 1. Two schematic metamorphic models to interpret temperatures: Temperatures are reset during retrograde cooling of a thermal event M1 (upper half), temperatures record a second lower grade metamorphic event M2 (lower half).

The following example gives a more detailed picture of the problems which have to be considered when interpreting isotope temperatures. In iron-rich ores from the Iron Quadrangle in Minas Gerais, Brazil, temperature information can be obtained only by analyzing the oxygen isotope composition of iron oxides (either hematite or magnetite) and of coexisting quartz. The area studied can be divided into two different regions: a western (W) region of greenschist assemblages and an eastern (E) region of amphibolite facies with transitions into granulite facies assemblages. As Fig. 2 shows, oxygen isotope fractionations (and therefore temperatures too) are obviously related to the deformation of the iron ores. Samples with a primary schistosity (S_1) only give the highest temperatures. In the high metamorphic E-region, temperatures vary between 815 and 635° C (upper curve) or 675 to 500° C (lower curve), depending upon which calibration curves are used (more details in HOEFS et al., 1982, and MÜLLER, et al., 1982). In the W-region with lower metamorphic grade S_1-samples cover the temperature range between 590 and 460° C (upper curve), 470 and 380° C (lower curve). Samples which have been affected by later deformation events and show signs

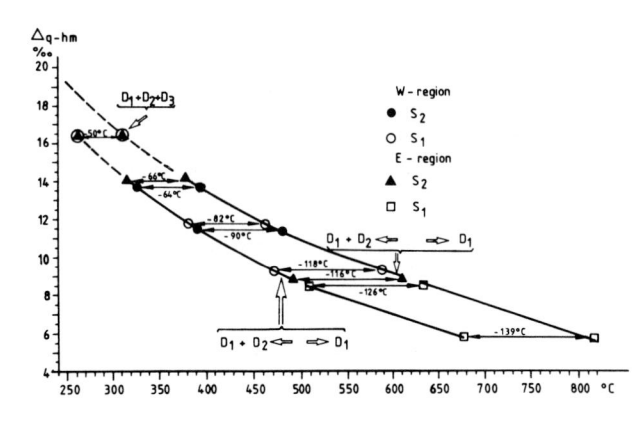

Fig. 2. Quartz (q)-hematite (hm) fractionations versus calculated temperatures (after MÜLLER et al., 1982). The different curves have been constructed according to different laboratory calibrations (for more details, see also HOEFS et al., 1982 and MÜLLER et al., 1982). D_1, D_2 and D_3 denote different deformation events.

of a schistosity S_2 and in a few cases $S_2 + S_3$, are selectively reset to lower isotopic temperatures. The more closely spaced the schistosity planes, the larger the extent of a temperature lowering.

Sulfur

A number of experimental studies have been carried out during the last 15 years. As with oxygen, the agreement of the experimental determinations is not very satisfactory. Two approaches have been used: one keeps both sulfides physically separated in the equilibrium vessel and effects isotope exchange via transport of sulfur vapor, the other uses hydrothermal solutions instead of a gas phase. OHMOTO & RYE (1979) examined critically all the available experimental data in terms of

a) attainment of equilibrium
b) uncertainties in the measurements
c) minimum or maximum fractionation factors when equilibrium was not attained
d) compatibility with the fractionation factors estimated from other sets of experiments.

Table 2 gives a summary of temperature coefficients A for some important sulfide minerals, which according to OHMOTO & RYE (1979) represent the best available numbers.

Table 2. Coefficients A for geological important sulfide pair fractionations (after OHMOTO & RYE, 1980)

	Sph	Pyrh	Cha	Ga
Py	0.3	0.3	0.45	1.03
Sph	-	0.0	0.15	0.73
Pyrh	-	-	0.15	0.73
Cha	-	-	-	0.58

Abbreviations: Py - pyrite, Sph - sphalerite, Pyrh - pyrrhotite, Cha - chalcopyrite, Ga - galena

Maybe the most suitable pair for calculating temperatures of mineralization is the sphalerite-galena pair. In Fig. 3 temperatures obtained from the sphalerite-galena pair are plotted against temperatures obtained from fluid inclusion studies; data are from the Providencia mine (RYE, 1974) and from the Finlandia Vein, Peru (KAMILLA & OHMOTO, 1977). The agreement between both temperatures is rather promising, however, the relatively large spread of the isotope temperatures is also noteworthy. Apparently, the greatest difficulty with respect to sulfide geothermometry lies in the fact that contemporaneous phases have to be selected for isotope measurements. It is obviously very difficult to obtain truely coexistent mineral pairs, so that both sulfides represent the same period of time in the history of the hydrothermal

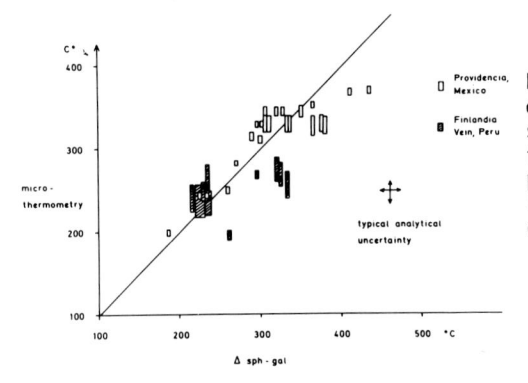

Fig. 3. Comparison of temperatures obtained by microthermometry (fluid inclusion studies) and by sulfur isotope fractionations of the sphalerite-galena pair. Data are from the Providencia mine (RYE, 1984) and from the Finlandia Vein deposit (KAMILLI & OHMOTO, 1977).

fluids. Even where a mineral pair belongs to the same period of mineralization, the isotopic composition and the temperature of the fluids may have been so variable that a relevant temperature information cannot be obtained. The best information may be obtained from those minerals which grew in contact with each other. Mineral pairs involving pyrite are less suitable for a temperature determination, because in many cases pyrite seems to precipitate over even larger portions of ore deposition than other sulfide minerals.

Hydrogen

OH-bearing minerals, such as biotite, muscovite, hornblende, chlorite show constant fractionations versus temperatures amongst each other and therefore cannot be used as geothermometers (SUZUOKI & EPSTEIN, 1976). In addition, it is difficult to establish whether hydrogen isotope equilibrium is commonly preserved between hydrous minerals and the temperatures at which cessation of hydrogen isotope exchange occurs.

Carbon

Besides the gaseous species CO_2 and CH_4 the calcite-graphite fractionation has a suitable magnitude to be applied for geothermometry. Studies by VALLEY & O'NEIL (1981) and WADA & SUZUOKI (1983) showed that this mineral pair can be successfully applied in carbonate-rich high metamorphic rocks at temperatures of above about 550° C. At temperatures below 550° C, kinetics of exchange may become so important that a general application of this thermometer is hazardous. The application of the calcite-graphite fractionations as a suitable thermometer has been questioned, however, just recently by KREULEN and van BEEK (1983).

Conclusions

Since isotope exchange reactions are almost completely independent of pressure, isotope geothermometers might appear to be the ideal thermometers, especially for metamorphic rocks, where isotope fractionations are reasonably large enough to give temperatures with relatively small errors. However, several problems complicate this simple picture. Firstly, there is a lack of calibration of some systems and - as shown above - disagreement in the available calibrations for some minerals. With the recently published set of mineral pair fractionations for oxygen (MATTHEWS et al., 1983a,1983b) this problem is on the way to be solved, at least for oxygen. Secondly, temperatures obtained by isotope fractionations are often difficult to interpret: sometimes they are in favour of peak metamorphic conditions; sometimes they indicate retrograde isotope exchange during the cooling period of a high temperature terrain. A distinction between these two temperature records might be possible when a comparison with other geological thermometers is carried out. In this respect more studies combining isotope thermometers with other geological thermometers, such as those by DAHL (1979) and GOLDMAN & ALBEE (1977), are required and will, certainly, give much more information about the complex temperature history of a rock.

As far as the geothermal systems are concerned, although there are potentially many isotope exchange processes occurring within a geothermal fluid, only a few have been generally applied, because of a suitable rate of achieving isotopic equilibrium. Such exchange reactions achieving equilibrium at different rates should be capable of indicating temperatures at various depths within a geothermal system. This approach is based on the assumption that during the ascent of the geothermal fluid to the surface a series of isotope equilibria are set up, and as the temperature falls, the rate of re-establishment of equilibrium is reduced until, finally, it exceeds the time taken for the fluid to reach the surface. One such example is shown in Table 3 for the geothermal fluids at Wairakei, New Zealand (HULSTON, 1976).

Table 3. Isotope temperature and rates of exchange to establish equilibrium for the hydrothermal fluid at Wairakei, New Zealand (HULSTON, 1976)

Element	Species	Isotope temperature	Rates of exchange
C	$^{13}CH_4 - ^{12}CO_2$	350° C	$10^2 - 10^5$ y
S	$H^{34}SO_4 - H_2^{32}S$	350° C	10^3 y
O	$HS^{16}O_4 - H_2^{18}O$	280° C	1 y
H	$H_2 - HOO$	260° C	1 - 2 weeks

Drillhole temperature 260° C

RELATIONS BETWEEN COALIFICATION AND PALAEOGEOTHERMICS IN VARISCAN AND ALPIDIC FOREDEEPS OF WESTERN EUROPE

TEICHMÜLLER, R.[†] and M. TEICHMÜLLER
Geologisches Landesamt Nordrhein-Westfalen
de Greiff Str. 195, D-4150 Krefeld, F.R. of Germany

Abstract

The degree of coalification ("rank") which may be measured as vitrinite reflectance (% Rm) under the microscope in almost all sedimentary rocks depends mainly on the maximum rock temperature - to a minor degree on the "cooking time". It is for this reason that close relationships exist between coalification and palaeogeothermics. The degree of rank increases with depth (the "rank gradient") commonly reflects the maximum geothermal gradient which was active in a certain profile and area. Palaeogeothermal gradients may be estimated from coalification gradients (measured in deep boreholes) if the burial history (m/Ma) is known.

In the subvariscan foredeep of the Ruhr Basin, coalification was completed before the Permian, due to the Asturian folding and uplift. The coalification gradients are one order of magnitude higher (0,5 - 1,0 % Rm/km) than in the foredeeps of the northern Alps and the northern Apennines (0,03 - 0,09 % Rm/km), although the depths of burial and the duration of heat exposure were similar. The reasons for the great difference are significantly higher geothermal gradients in the Ruhr Basin during the Upper Carboniferous. According to BUNTEBARTH et al. (1982) these gradients vary between 60 and 80° C/km, whereas in the foredeep molasses of the northern Alps and the northern Apennines they range between 19 and 23° C/km due to subduction and the resulting increase of crustal thickness. The high geothermal gradients of the Subvariscan foredeep (60 - 80° C/km against the present 30° C/km) suggest a thinner crust during the Carboniferous. Based on the correlation between heat flow density and crustal thickness reported by ČERMÁK (1979), values of 110 - 130 mW/m², estimated for the Ruhr Carboniferous, correspond to 20 - 22 km crustal thickness. Thus results of coalification studies agree with results that ZWART (1967,1976) obtained from studies of the regional metamorphism of rocks within the Variscan orogen of Europe.

Fundamentals of relations between coalification and geothermics

Organic substances in sediments are very sensitive to heat exposure. Normally they are coalified with increasing depth of burial, due to increasing rock temperature. The degree of coalification ("rank") depends mainly on the maximum temperature and

- to a minor degree - on the heating time. In contrast to transformations or neoformations of minerals, which are also used as geologic thermometers, the degree of coalification is not influenced by such conditions as pH, eH, partial pressure of water, or ion concentrations, and moreover, it is irreversible. It can me measured by optical reflectance of the coal maceral vitrinite which is derived from humufied plant material, in particular from lignin and cellulose. Even tiny particles of vitrinite (>3 µm), which occur in almost all clastic rocks, suffice to determine the rank. The reflectance is measured on the surface of polished sections of coal or rock with the help of a microscope photometer (for method see STACH et al., 1982). Fig. 1 shows particles of vitrinite in a siltstone, under the microscope.

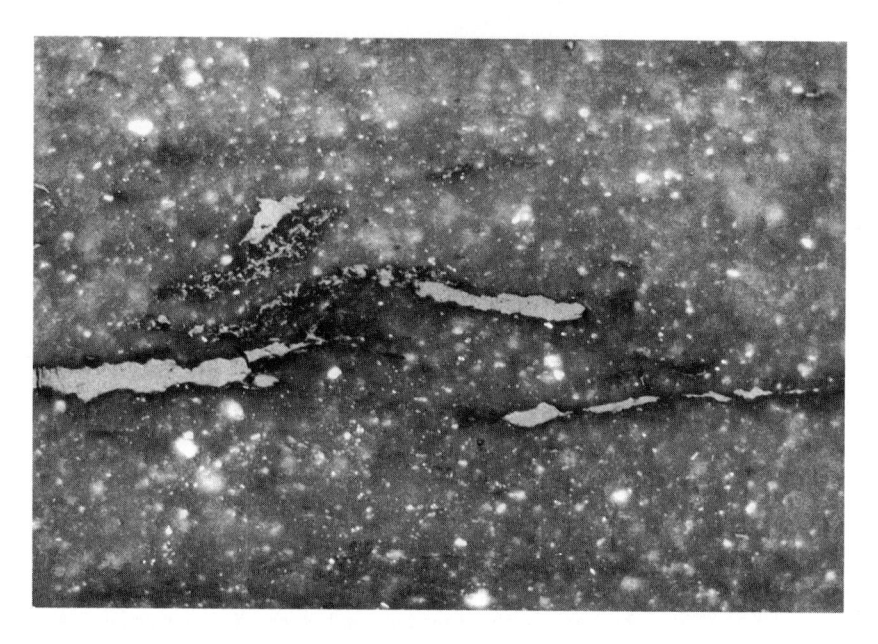

Fig. 1. Vitrinite inclusions in siltstone. Photomicrograph of polished section, oil immersion, 500 x magnified.

Many authors have developed empirical and/or hypothetical diagrams and formulars showing the relations between vitrinite reflectance, rock temperature and "cooking time", - e.g. LOPATIN (1971), LOPATIN & BOSTICK (1973/74), HOOD et al. (1975), TISSOT & ESPITALIÉ (1975), BURNE & KANSTLER (1977), BUNTEBARTH (1978,1979) and WAPLES (1980). Fig. 2 shows a diagram from BOSTICK et al. (1979). The "effective heating time" (HOOD et al., 1975) is the time during which the coal was within 15° C of its maximum burial temperature.

Since vitrinite reflectance is related to the degree of aromatization of the humic complexes in coal (McCARTNEY & M. TEICHMÜLLER, 1972) and since aromatization runs reciprocal to the volatile matter of vitrinite (as a main chemical rank parameter for

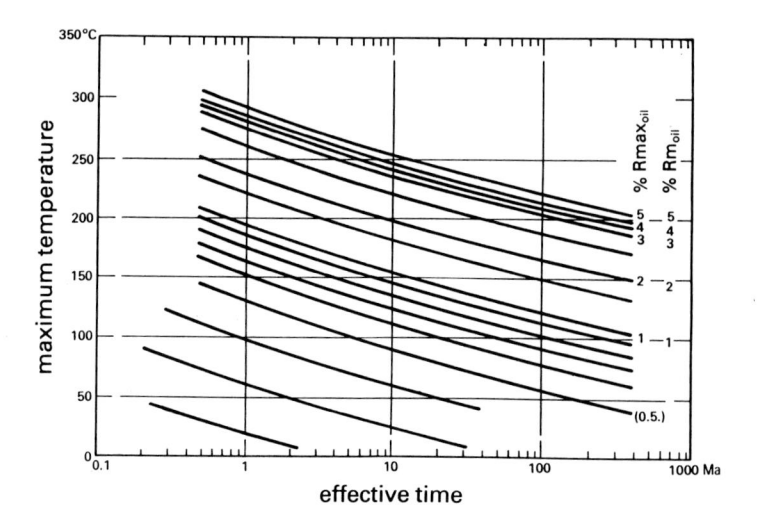

Fig. 2. Relationship between maximum rock temperature, effective heating time and vitrinite reflectance (Rm, Rmax) (after BOSTICK et al., 1979).

coal), Fig. 3 demonstrates the different degrees of increase of vitrinite reflectance in the different rank stages: vitrinite reflectance rises only slowly in the low rank stages (brown coal, lignite, high volatile bituminous coal), more rapidly in the stage of medium and low volatile bituminous coals, and particularly strongly in the stage of anthracites. The degree of rank increases with depth, i.e. the "rank gradient", depends not only on the rank stage but also, usually on the geothermal gradient active during coalification. In Fig. 4 rank increase is shown for four boreholes which encountered young sediments of the Upper Rhine Graben - one with a geothermal gradient of 42°C/km, the others with gradients ranging from 67° C/km to 77° C/km. The more rapid increase of vitrinite reflectance in the "warm" boreholes is evident. Fig. 5 is a corresponding plot of Rm-increase against the present rock temperature in the same boreholes. The good correlation between Rm and temperature (as opposed to the bad correlation between Rm and depth in Fig. 4) shows the strong influence of the geothermal gradient upon rank increase.

Rank gradients measured in deep boreholes allow estimations of palaeogeothermal gradients if the burial history is known (e.g. ESPITALIE, 1979; BUNTEBARTH, 1979). Fig. 6 gives an example for the burial history (meter per million years) of the Lower Oligocene in a borehole of the Upper Rhine Graben. Based on rank gradients from deep boreholes and corresponding burial histories, we suggested two heat maxima for the middle Upper Rhine Graben, - one during the early Tertiary and the other, which is still active, since the Upper Pliocene (Fig. 7).

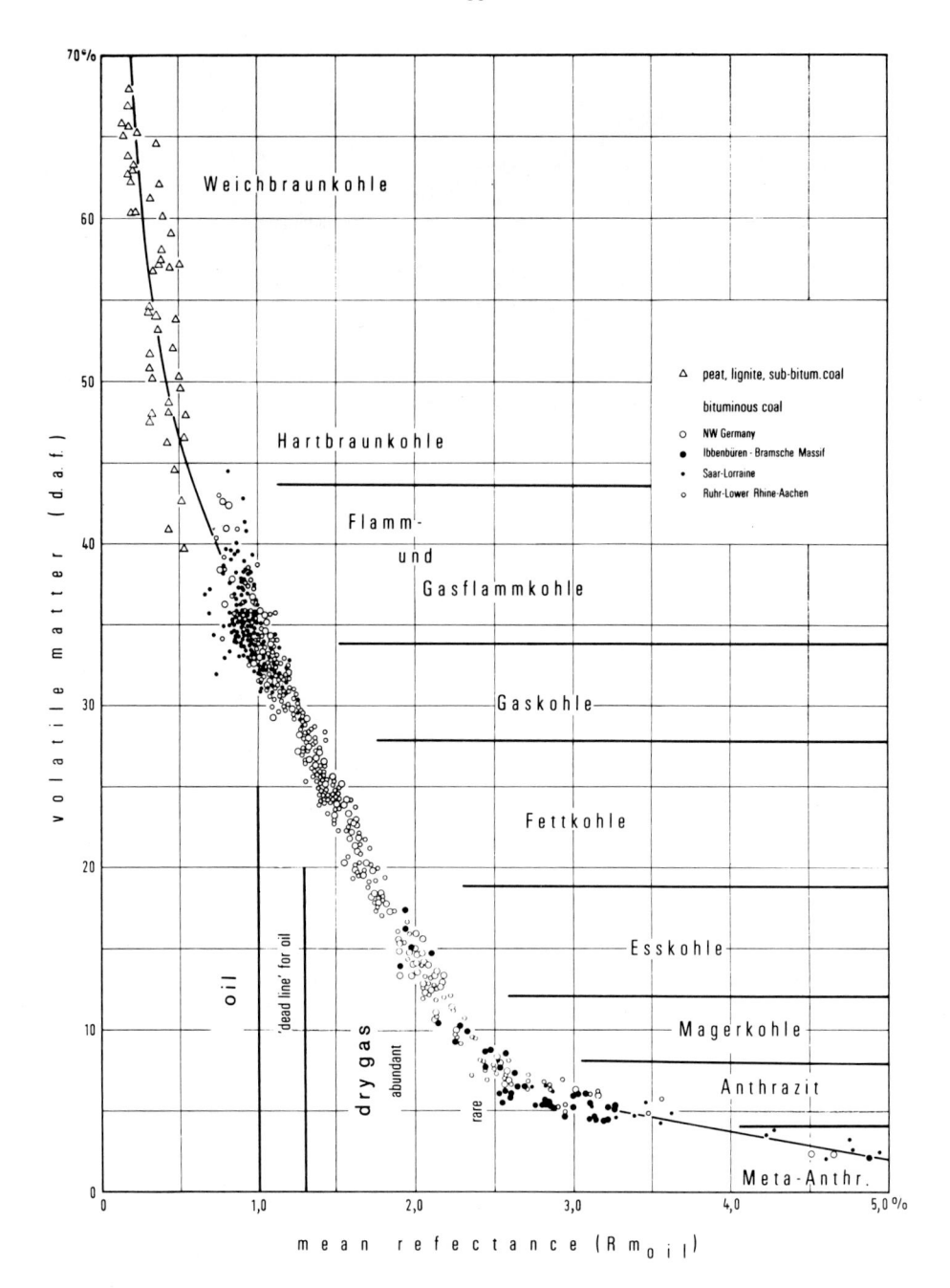

Fig. 3. Relationship between vitrinite reflectance and volatile matter of vitrites for the different rank stages (from soft brown coal to meta-anthracite) of German coal deposits (after M. TEICHMÜLLER & R. TEICHMÜLLER, 1979b).

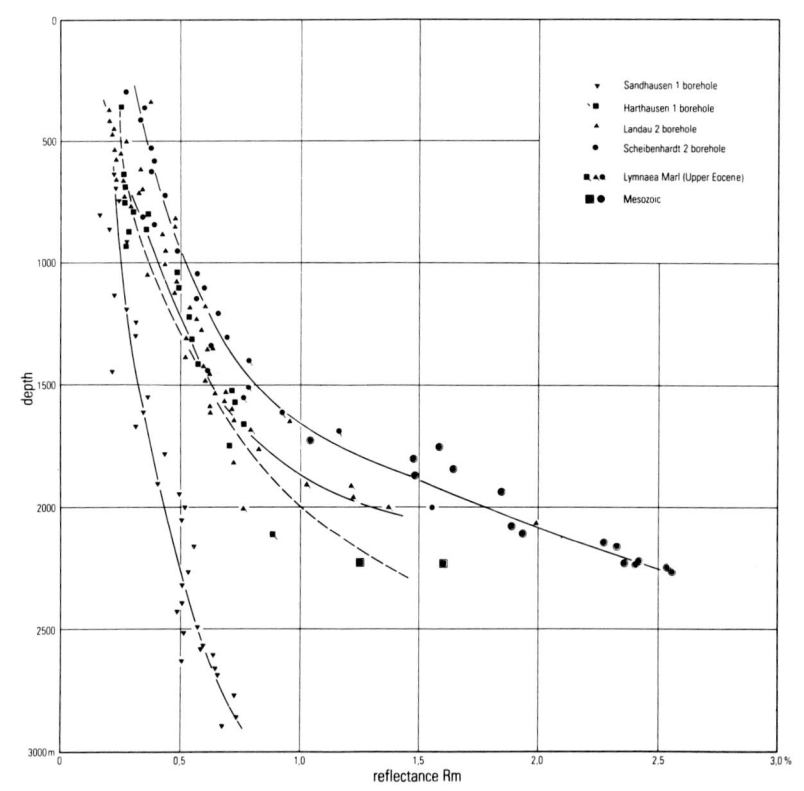

Fig. 4. Relationship between vitrinite reflectance and depth in four boreholes of the Upper Rhine Graben. The geothermal gradients are: 40° C/km for Sandhausen borehole and 70 - 80° C/km for the remaining boreholes (after M. TEICHMÜLLER & R. TEICHMÜLLER, 1981). Compare with Fig. 5!

In the Saar District of Western Germany, the very deep borehole Saar 1 encountered coal seams of the Upper Carboniferous, the vitrinite reflectance of which was determined between 410 and 4470 m depth. The results are shown in Fig. 8, together with results from some other boreholes of the Saar-Nahe Basin. The present temperature gradient in the Saar 1 borehole is 29° C/km (HEDEMANN, 1976). According to computation by BUNTEBARTH (1983) based on the rank of coal and the burial history, the palaeogeothermal gradients varied between 85 and 55° C/km in Permo-Carboniferous times. A maximum coalification temperature of 350° C for the Westphalian A (present depth: 4470 m) was also calculated and the result is in agreement with the first occurrence of the mineral epidot at a depth of 5037 m in the Upper Devonian of the Saar 1 borehole (ZIMMERLE, 1976). The boundary between diagenesis and anchimetamorphism, based on thorough studies by BREITSCHMID (1982) in the Swiss Alps, corresponds to a rock temperature of 220° C, and has been encountered in the Saar 1 borehole at a depth of 3470 m. The present temperature at this depth is only 95° C. These findings suggest a much higher geothermal gradient during the Permo-Carboniferous in the Saar District than is measured at present. The Saar coal is part of the backdeep

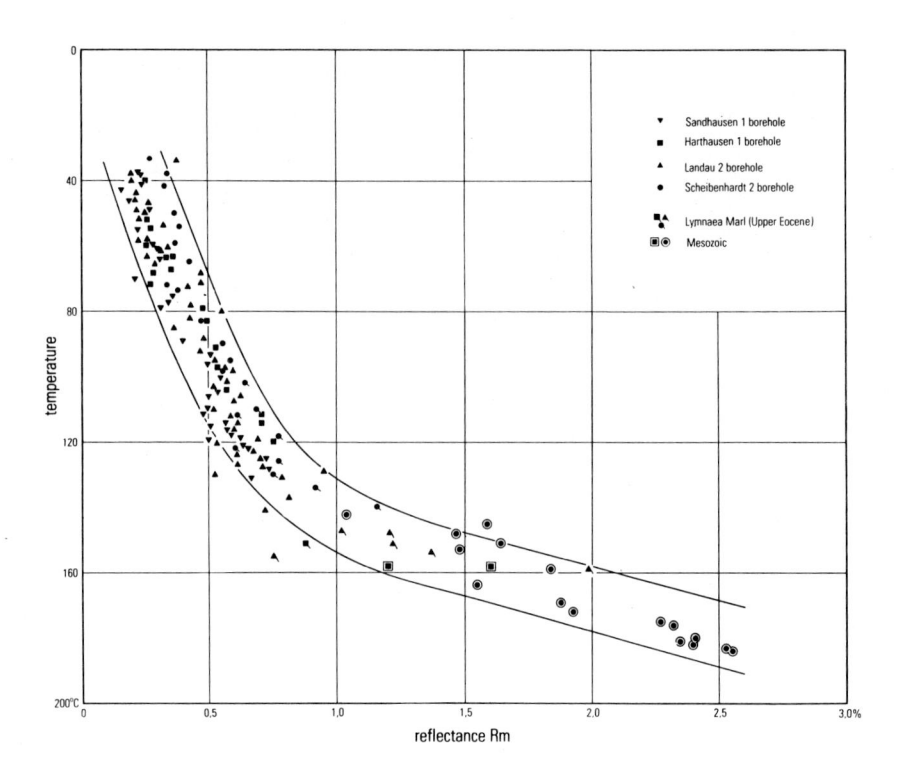

Fig. 5. Relationship between vitrinite reflectance and rock temperature in four boreholes of the Upper Rhine Graben (same boreholes as in Fig. 4) (after M. TEICH-MÜLLER & R. TEICHMÜLLER, 1981).

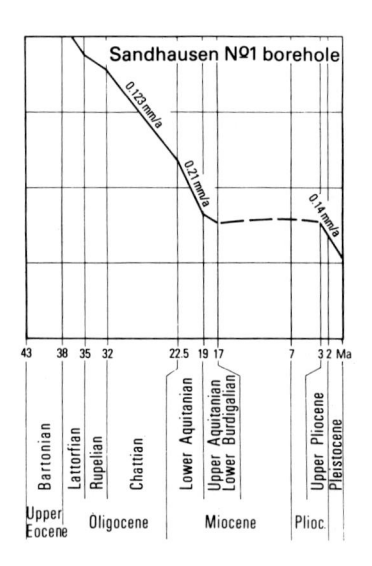

Fig. 6. Subsidence curve for the Lower Oligocene in the Sandhausen 1 borehole (Upper Rhine Graben) (after M. TEICHMÜLLER & R. TEICHMÜLLER, 1979).

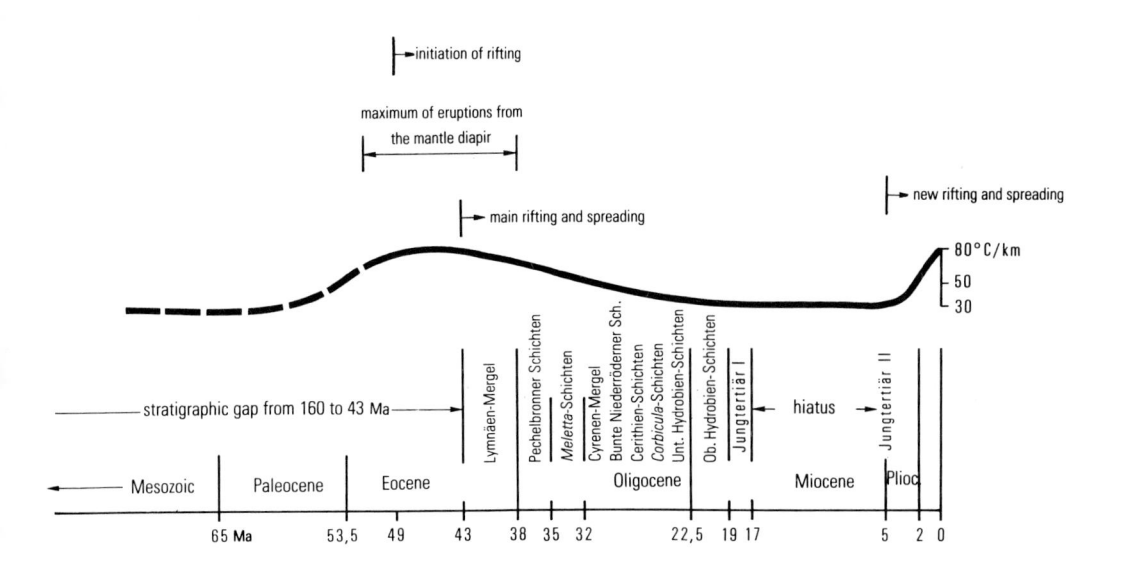

Fig. 7. Variation of geothermal gradients during the development of the middle Upper Rhine Graben, deduced from results of coalification studies (after M. TEICH-MÜLLER & R. TEICHMÜLLER, 1979).

of the Rhenoherzynian Variscan Mountains. Similar high geothermal gradients have been reported from the Ruhr Basin which, in contrast to the Saar Basin, belongs to the Subvariscan foredeep.

Foredeeps are often rich in coal, oil and gas deposits. Therefore, many deep boreholes are commonly put down in foredeeps. Coalification studies in these boreholes allow certain reconstructions of palaeogeothermics. In the present paper, results of coalification studies and their geothermal evaluations for the Subvariscan foredeep of the Ruhr Basin and the Alpidic foredeeps of the northern Alps and northern Apennines are reported.

Palaeogeothermics of the Subvariscan foredeep (Ruhr Basin and Lower Rhine Basin)

Due to many exposures in mines, quarries and, particularly, in deep boreholes of the Ruhr industry, the coalification pattern of the Ruhr and Lower Rhine Basin are well known (M. TEICHMÜLLER & R. TEICHMÜLLER, 1949,1968,1971; PATTEISKY et al., 1962). The rise of vitrinite reflectance with depth has been studied in many boreholes by M. TEICHMÜLLER & M. WOLF (Geologisches Landesamt Nordrhein-Westfalen, Krefeld) during a period of more than twenty years, so that the rank gradients could be evaluated for palaeogeothermal studies by KOPPE (1980) and BUNTEBARTH et al. (1982a). The evaluations were facilitated by the fact that, at the Ruhr, coalification took place in Carboniferous times before the Asturian folding occurring in Late Westphalian D and Stephanian times.

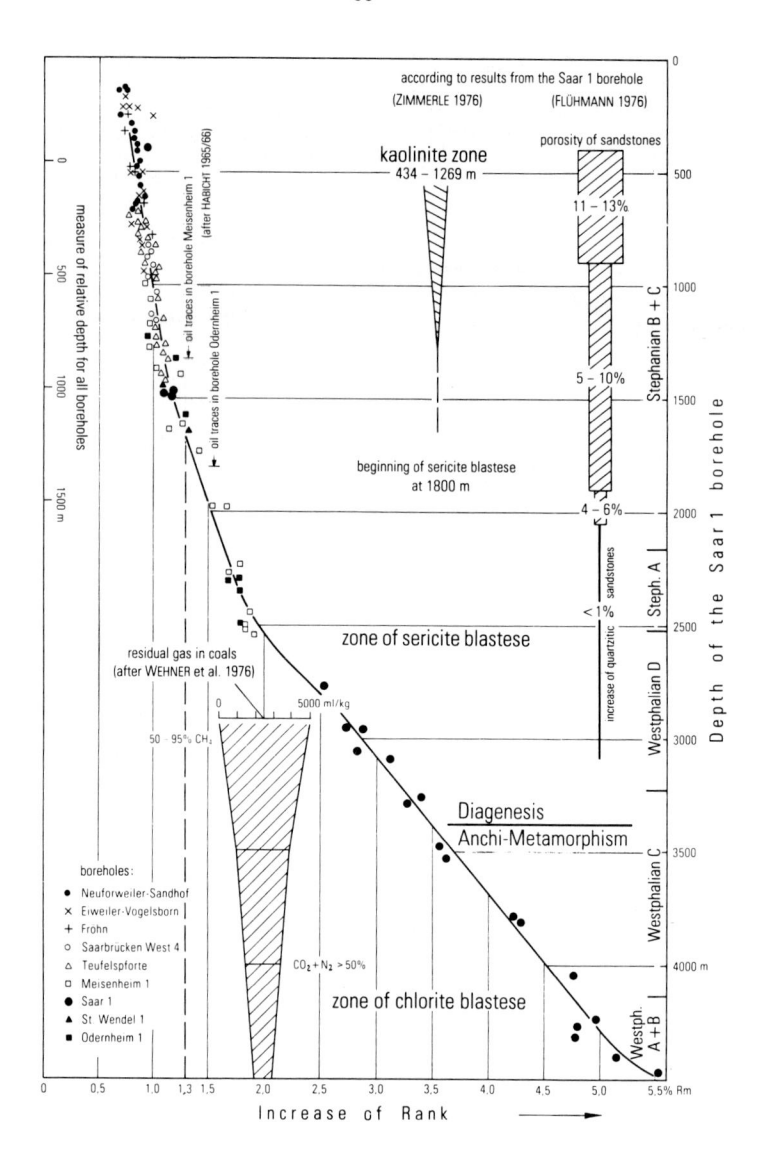

Fig. 8. Increase of vitrinite reflectance with depth in the Saar 1 borehole and in some other boreholes of the Saar-Nahe Basin (West-Germany). The ordinate to the left indicates relative depth and is valid for all boreholes, whereas the ordinate to the right is valid for the Saar 1 borehole only (after M. TEICHMÜLLER et al., 1983).

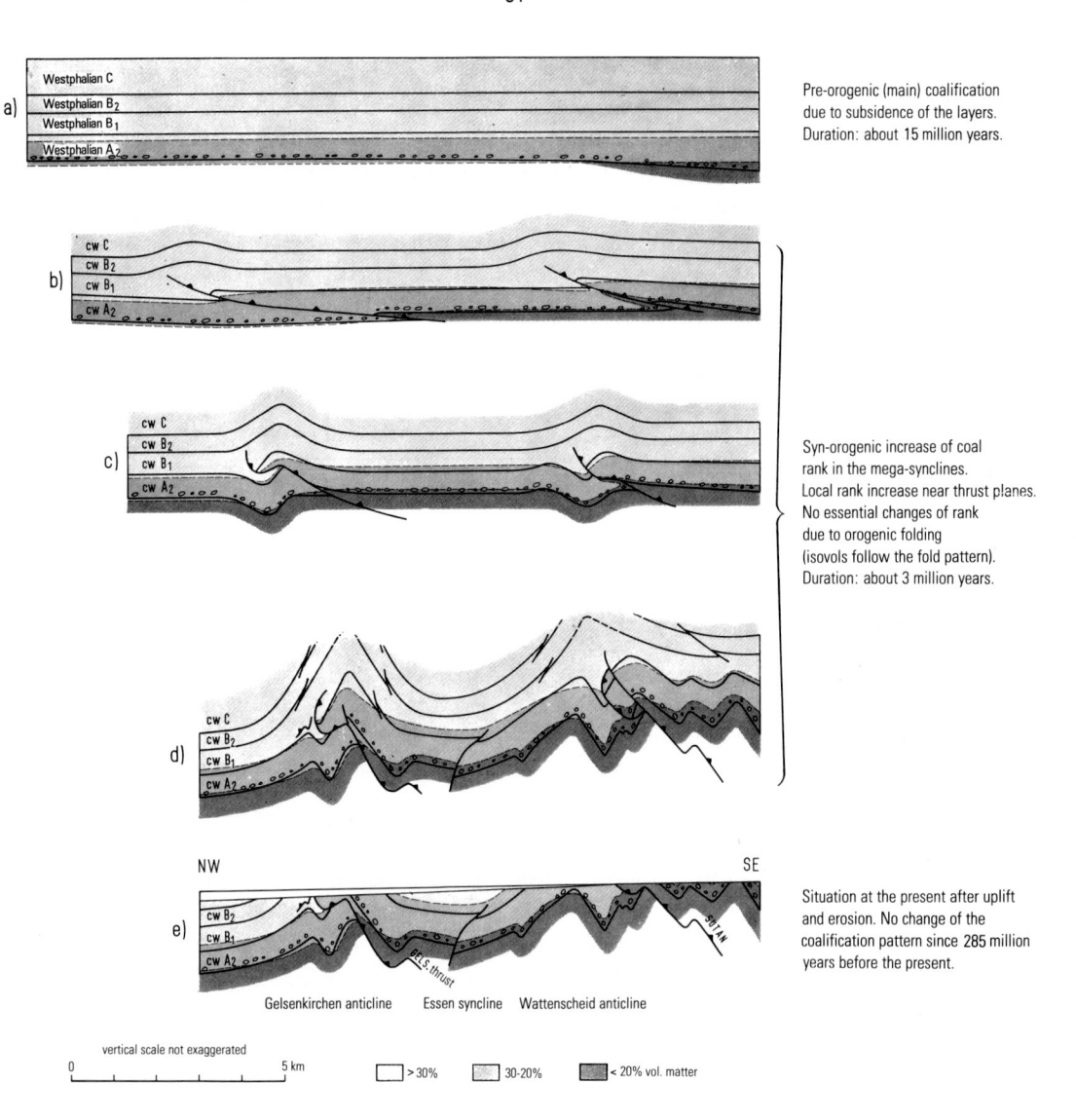

Fig. 9. History of coalification in the Ruhr Carboniferous (after M. TEICHMÜLLER & R. TEICHMÜLLER, 1971).

Fig. 9 shows the development of coalification in relation to thrusting and folding. Zones of equal rank (volatile matter) run parallel to the stratification. After folding, the isorank lines follow the fold pattern. Thus, coalification is pre-orogenic. This also becomes evident by very low rank Permian (Kupferschiefer in the brown coal stage) overlying strongly coalified Westphalian. In the Ruhr Basin the Cretaceous cover was not thick enough to promote the Carboniferous coalification. Similar conditions prevailed on the western side of the Rhine. Fig. 10 demonstrates the typical subsidence/uplift history for a coal seam in the Lower Rhine Basin.

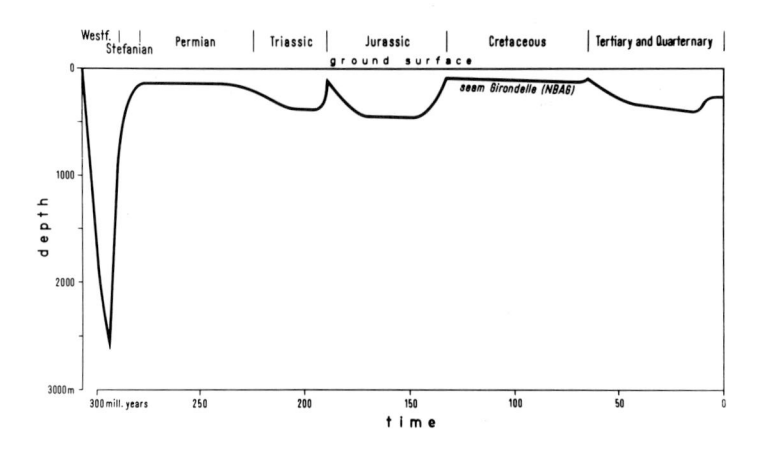

Fig. 10. Burial history of Girondelle Seam (Westphalian A) in the Lower Rhine Basin.

Already in former times coalification studies were used to estimate palaeogeothermal gradients for the Ruhr Carboniferous. A comparison of subsidence curves for the early Tertiary of the Upper Rhine Graben and for the Ruhr Carboniferous (cf. Fig. 17), taking into consideration the temperatures measured in boreholes of the Upper Rhine Graben (cf. Fig. 5), led to the conclusion that in the Ruhr Basin the palaeogeothermal gradients reached 50 - 70° C/km during the Westphalian (R. TEICHMÜLLER, 1973).

Later, KOPPE (1980) and BUNTEBARTH et al. (1982a) evaluated more than 1200 data of vitrinite reflectance obtained by M. TEICHMÜLLER and M. WOLF from 44 deep boreholes of the Ruhr and Lower Rhine area. As in the Saar 1 borehole, the rank gradients in boreholes of the Ruhr and Lower Rhine Basin are relatively high. According to KOPPE (1980) and BUNTEBARTH et al. (1982a), they vary, on average, between 0,48 % Rm/km for the Westphalian C and B and 1,04 % Rm/km for the Westphalian A. Extreme values are 0,20 % Rm/km and 1,4 % Rm/km. BUNTEBARTH et al. (1982a) evaluated the coalification gradients of more than 50 boreholes to estimate the palaeogeothermal gradients for the Upper Carboniferous at the Ruhr. Fig. 11 shows the result of these calculations

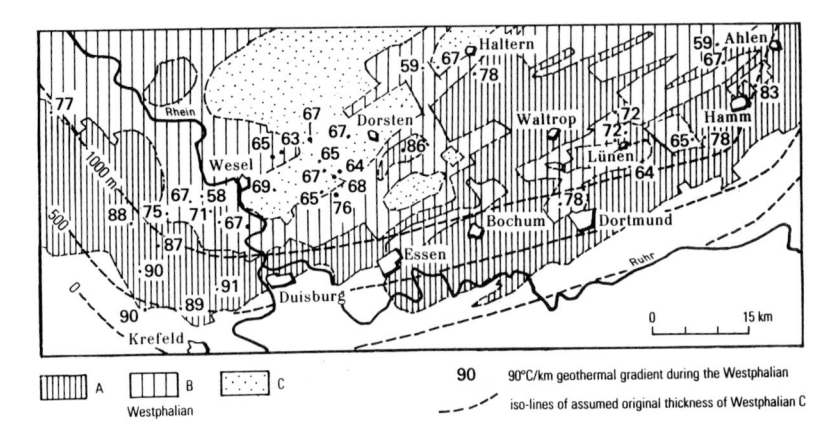

Fig. 11. Palaeogeothermal gradients during the Upper Carboniferous (Westphalian A, B, C) in the Ruhr Basin and the adjacent Lower Rhine Basin of West-Germany. The gradients were computed on the basis of 1600 reflectance data from 53 deep boreholes (after BUNTEBARTH et al., 1982a).

based on rank gradients, depth and duration of burial. The duration of burial between sedimentation and upfolding was about 15 million years for a seam like Katharina at the boundary between Westphalian A and B. The maximal burial was about 2500 m, the effective heating time perhaps 2 million years.

According to Fig. 11 the computed geothermal gradients commonly vary between 60 to 70° C/km. Locally gradients of 80 - 90° C/km were computed. These high gradients have been explained by deep lying crypto-plutons north of Krefeld (91° C/km) and near Westerholt (86° C/km). The Krefeld pluton intruded in Permo-Carboniferous times as suggested by the occurrence of diabase dykes cutting Westphalian strata. Moreover, a magnetic anomaly and the occurrence of very strongly coalified coals are arguments for an intrusive body (BUNTEBARTH et al., 1982b). Near Westerholt, in an area with high coalification gradients, an unusual occurrence of newly formed sericite in the Westphalian A (WEGEHAUPT, 1962) as well as ore mineralization (STADLER, personal communication) and special structural indications (ADLER, 1961) suggest a magmatic intrusion at depth as an additional heat source.

North of the Ruhr Basin, the Cretaceous cover increases in thickness. The Münsterland 1 borehole encountered 3650 m of Upper Carboniferous below a cover of 1788 m Cretaceous. As in the Saar 1 borehole (Fig. 8), in the Münsterland 1 borehole the rank gradient (vitrinite reflectance) rises with increasing depth and increasing rank range (Fig. 12). In the Westphalian B and A the gradient is 1,05 % Rm/km, whereas in the Namurian it amounts to 1,75 % Rm/km.

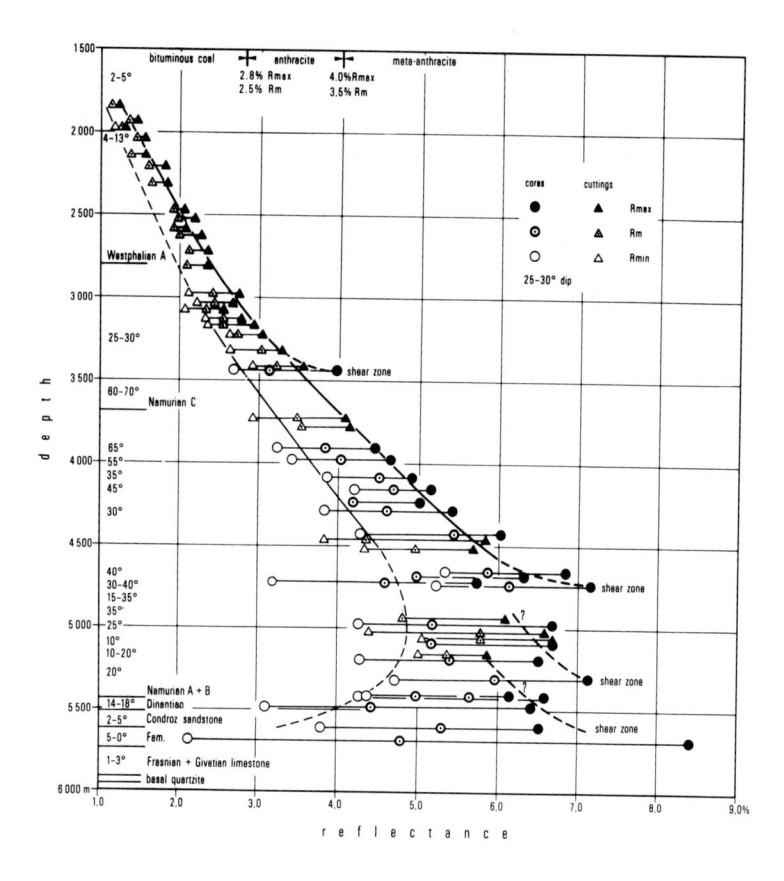

Fig. 12. Increase of vitrinite reflectance (% Rmax, Rm and Rmin) with depth in the Münsterland 1 borehole. Note the increase of anisotropy (difference between Rmax and Rmin) with increasing depth and rank stage (after M. TEICHMÜLLER, 1982).

Taking the burial histories into consideration, G. BUNTEBARTH kindly computed palaeogeothermal gradients for some deep boreholes north of the Ruhr Basin (Table 1). The gradients lie in the same order of magnitude as those computed for the Saar 1 borehole (BUNTEBARTH, 1983). According to Table 1 it seems that the gradients have been still higher in the Namurian than in the Westphalian.

Fig. 13 shows the remarkable coincidence of rank gradients in the same deep boreholes north of the Ruhr Basin.

Earlier suggestions of high geothermal gradients in the Variscan foredeep of western Germany, were based on the following: In the borehole Münsterland 1 the present depths correspond more or less to the maximal depths of the layers in Carboniferous times. According to cathode-luminescent properties of quartz in Devonian quartzites, ZINKERNAGEL (1978) suggests a palaeogeotemperature of >300° C (present temperature

Table 1. Rank gradients and geothermal gradients in deep boreholes north of the Ruhr Basin and in the Saar 1 borehole.

Borehole	Depth (m)	Stratigraphy	Rank Range (% Rm) according to Fig. 13	Rank Gradient (% Rm/km)	Geothermal Gradient (°C/km) palaeo*)	recent
Isselburg 3	1348-2130	Westphalian A	1.0 - 1.7	0.89	69	
	2180-3564	Namurian	1.7 - 3.62	1.39	78	not known
Münsterland 1	1843-2796	Westphalian B - A	1.18 - 2.18	1.05	80	
	2820-4700	Namurian	2.20 - 5.50	1.75	87	37
Versmold 1	1456-3202	Namurian	2.35 - 5.20	1.63	88	33
Saar 1	2772-4400	Westphalian D - B	2.48 - 5.25	1.70	85	29

*) computed by Dr. G. BUNTEBARTH

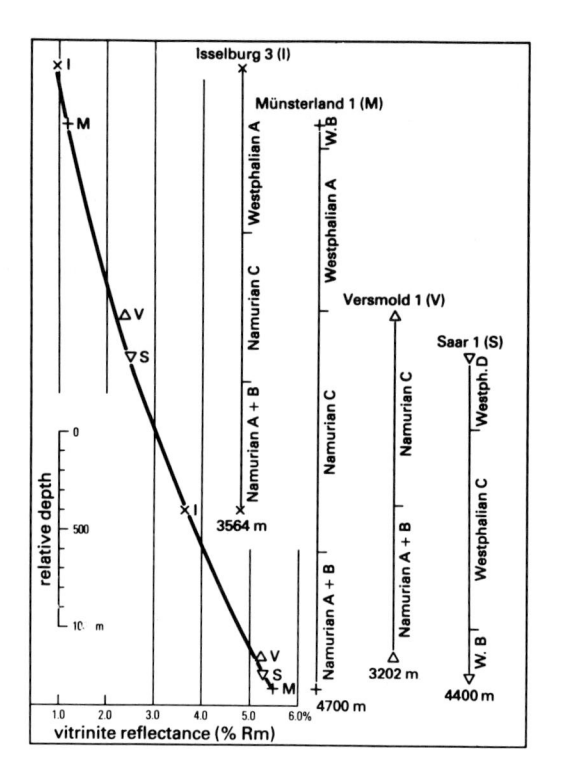

Fig. 13. Increase of vitrinite reflectance with depth in the deep boreholes Isselburg 3, Münsterland 1, Versmold 1 and Saar 1.

about 200° C) at a depth of 5700 m, corresponding to a palaeogeothermal gradient of >50° C/km.

STRASSER & WOLTERS (1963) determined a volumetric density of 2,8 g/cm³ for Devonian clayey sandstones in the Münsterland 1 borehole at 5700 m. After a diagram from CISSARZ (1965: Fig. 2) this value corresponds to a rock temperature of 300° C.

According to BREITSCHMID (1982) the top of the epi-zone (boundary between very low grade and low-grade metamorphism) corresponds to a vitrinite reflectance of 5,5 % Rm and to a rock temperature of 270° C in the Swiss Helveticum. In the Münsterland 1 borehole a vitrinite reflectance of 5,5 % Rm is reached at a depth of 4700 m (Fig. 12) which indicates a palaeogeothermal gradient of 57° C/km, facing a gradient of 33,6° C/km at the present time (HEDEMANN & R. TEICHMÜLLER, 1966).

A comparison of subsidence curves for the early Tertiary of the Upper Rhine Graben and for the Ruhr Carboniferous (Fig. 17), with the temperatures measured in boreholes of the Upper Rhine Graben taken into consideration (Fig. 5), led to the conclusion that, in the Ruhr Basin, the palaeogeothermal gradients reached 50 - 70° C/km during the Westphalian (R. TEICHMÜLLER, 1973).

Geothermics of Alpidic foredeeps
Northern Alps

The molasse of the Alpidic foredeep north of the Alps is comparable to the molasse of the Subvariscan foredeep north of the Rhenohercynicum. Depths of burial and effective coalification times are similar in both cases (Fig. 17). However, the coalification pattern is very different.

In the foredeep of the Alps, burial depths of 2,5 km (near Munich) to 5 km (in the Alps) were attained for the basis of the molasse (JACOB et al., 1982). Folding and uplift began 8 to 12 million years before the present, in Late Miocene to Pliocene times. As usual in foredeeps, folding moved from the mountains (12 Ma) to the foreland (8 Ma).

Fig. 14 shows a cross-section through the foreland and the northernmost part of the German Alps. In the north, the Alpidic foredeep is filled with autochtonous, flat lying Tertiary molasse sediments. Approaching the mountains in the south, the molasse is folded and imbricated. Still further south, in the Alps, this imbricated, subalpine molasse is overlain by imbricated strata of the Helveticum, Flysch and Austroalpine nappe piles.

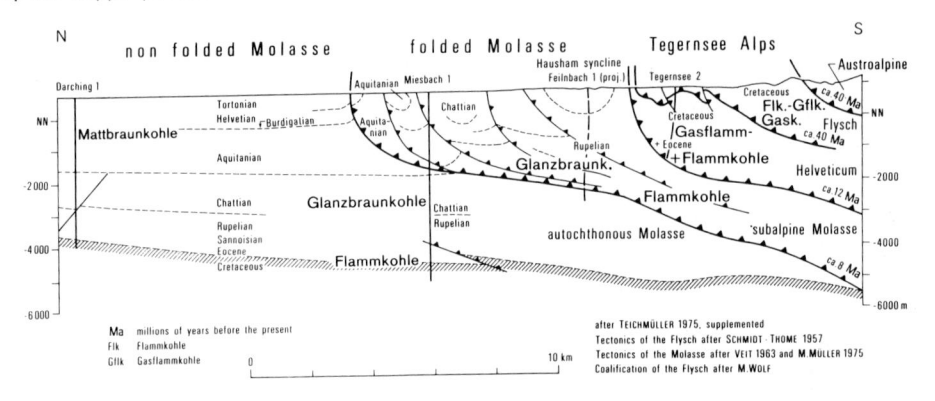

Fig. 14. Cross-section through the subduction zone at the northern rim of the Alps with coalification profiles (after M. TEICHMÜLLER & R. TEICHMÜLLER, 1978).

In the foreland as well as in the northern Alps, boreholes of the oil industry permitted a study of the downward increase of vitrinite reflectance. According to coalification studies of JACOB & KUCKELKORN (1977), the Miesbach 1 borehole did not encounter the boundary between sub-bituminous and bituminous coal (boundary between Braunkohle and Steinkohle according to the German coal classification) until a depth of about

5000 m in the molasse (see Fig. 14). In this borehole the present geothermal gradient is 23,5° C/km (JACOB & KUCKELKORN, 1977).

JACOB et al. (1982) evaluated coalification data from the Alpine molasse obtained by JACOB and other authors. They found that in the south, within the folded and imbricated molasse, the coalification gradients are especially low, and even too low compared with the present geothermal gradients. Therefore, these authors assume a prekinematic coalification having taken place 10 - 20 km further south where the temperature gradients are lower at present and obviously have been lower in pre-thrusting times as well. It seems that in this region the geothermal pattern of the northern Alps has not changed much since Pliocene/Upper Miocene times when the sediments of the present folded molasse reached their deepest level of subsidence and, thus, their highest coalification temperatures. In contrast to the northern Alps further west (e.g. near Oberstdorf, Allgäu and, particularly, in Switzerland) where the main coalification (as well as illite diagenesis) occurred after imbrication due to former thick overburden with Penninic nappes (M. TEICHMÜLLER & R. TEICHMÜLLER, 1978; FREY et al., 1973), at the northern rim of the Bavarian Alps no post-kinematic coalification of the folded and imbricated molasse took place. Fig. 15 shows the results of JACOB et al. (1982) with coalification gradients for the autochthonous as well as the folded and imbricated molasse in its pre-tectonic, non imbricated position. The gradual decrease from the Anzing 3 borehole in the north (lying 23 km north of the sign for Anzing 3 in Fig. 15) with 0,09 % Rm/km, to Hausham in the south with 0,04 % Rm/km, and further west, - from the Staffelsee 1 borehole in the north (0,06 % Rm/km) to the Egling 1 borehole in the south (0,03 % Rm/km), is evident. The present geothermal gradient in the Anzing 3 borehole in the north is 22,8° C/km, whereas it is 22° C/km in the Vorderriss 1 borehole (east of Garmisch-Partenkirchen) in the south (BACHMANN & MÜLLER, 1981).

The decrease of coalification gradients towards the south is obviously caused by a decrease of the geothermal gradients in the same direction, - not only in the present time but also during Late Miocene/Pliocene times when maximal depths of subsidence were reached (M. & R. TEICHMÜLLER, 1975; JACOB & KUCKELKORN, 1977; JACOB et al., 1982).

This geothermal pattern is in accordance with the crustal thickness increasing from north to south, due to subduction. Fig. 16 shows, over a horizontal distance of 55 km, the dipping of the Mohorovičić discontinuity from 33 km depth in the northern Alpine foreland (Anzing 3 borehole) down to 40 km depth beneath the northern Alps (Tegernsee 2 borehole). Thus, we must assume relationships between coalification gradients, geothermal gradients and the thickness of the crust in the foredeep of the German Alps.

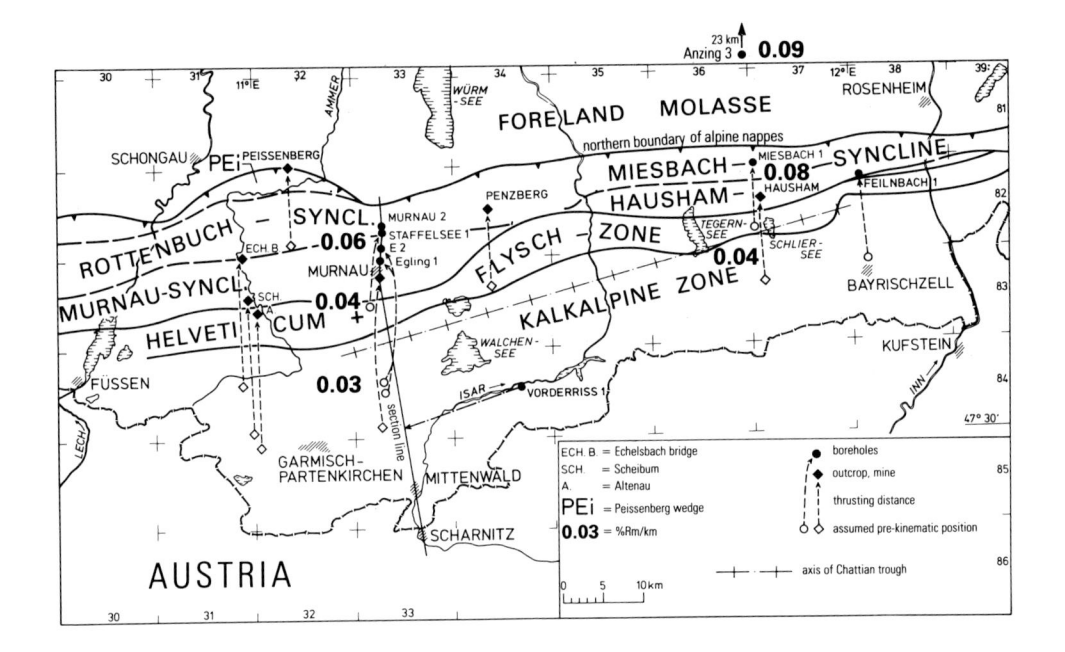

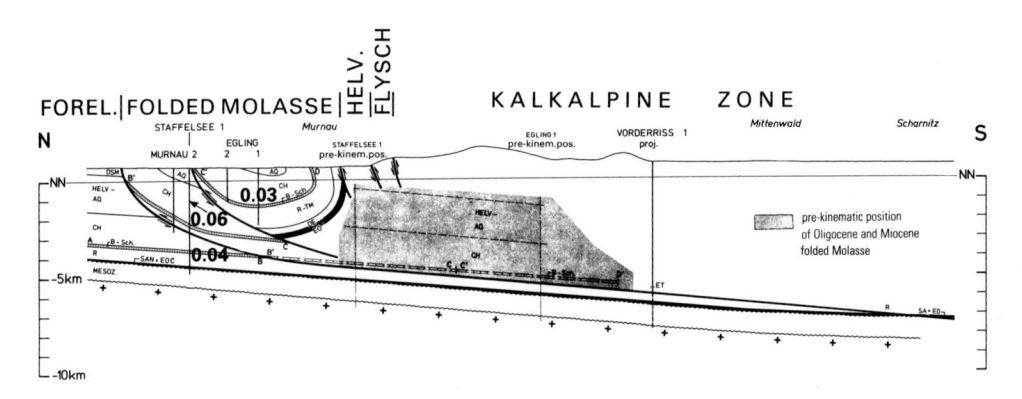

Fig. 15. Coalification gradients (% Rm/km) in the sub-Alpine molasse before dislocation of the folded molasse (after JACOB et al., 1982), coalification gradients introduced by the present authors. The map above shows the dislocation of the folded molasse between the rivers Lech and Inn. For position of the cross-section (below), see the map (above).

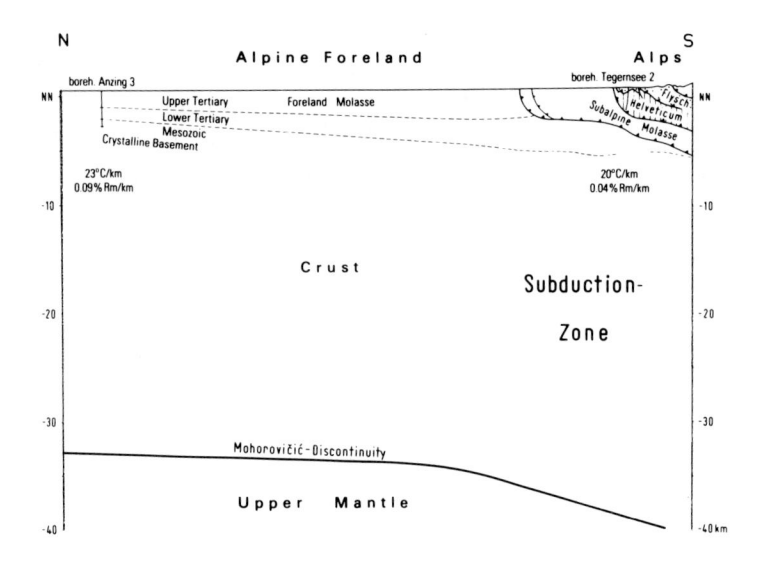

Fig. 16. The subduction zone at the northern margin of the Alps. Depth of boundary between crust and upper mantle according to GIESE & STEIN (1971).

It is interesting to note that in the sub-Alpine molasse the degree of coalification very seldom reaches the stage of bituminous coal, although depth of burial and effective coalification time were about the same as for the Ruhr bituminous coals. As a matter of fact, according to Fig. 17, the time to reach the maximum subsidence depth

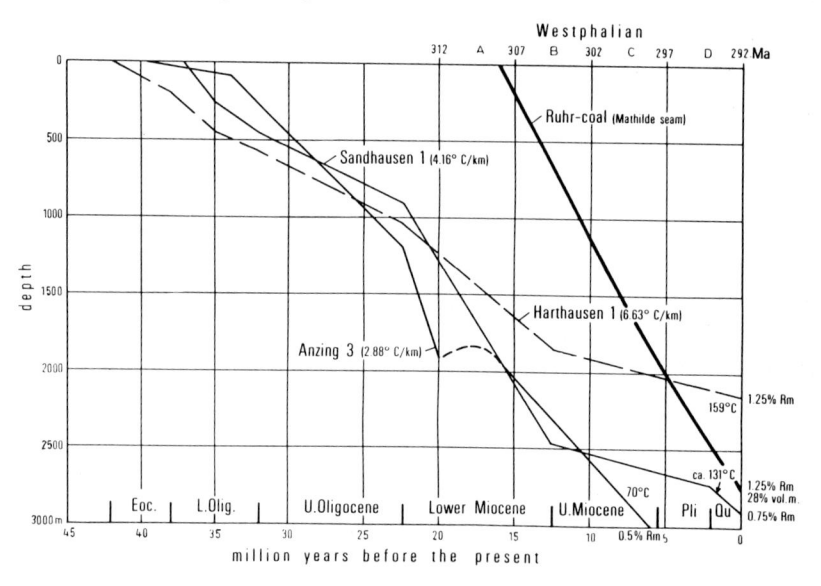

Fig. 17. Subsidence history, geothermal gradient and final rank of coal (% Rm) in the Anzing 3 borehole, compared with boreholes in the Upper Rhine Graben (Sandhausen 1 and Harthausen 1) and with a Ruhr coal (after R. TEICHMÜLLER, 1973; M. TEICHMÜLLER & R. TEICHMÜLLER, 1975).

and with that the maximum temperature was even longer for the Alpidic molasse than for the sub-Variscan molasse as may be seen from the subsidence curve for the Anzing 3 borehole compared with the curve for the Mathilde seam of the Ruhr Carboniferous.

Since the coalification gradients in the Alpidic molasse vary between 0,03 and 0,09 % Rm/km they are more than one order of magnitude lower than the coalification gradients of the Subvariscan molasse at the Ruhr (0,5 - 1,0 % Rm/km).

The explanation for this striking difference is the subduction zone beneath the northern Alps where the crust-mantle boundary lies as deep as 40 km and where thick piles of nappes are stacked one above the other (see Figs.14,16). In this respect, the conditions are similar to those at the northern margin of the Carpathian Mts. (SZADECZKY-KARDOSS, 1973), and to the subduction zone of western California where, according to BOSTICK (1974) and CASTANO & SPARKS (1974), geothermal gradients and coalification gradients are remarkably low.

Northern Apennines

The northern Apennines are a very young orogen in which imbrication of large nappes occurred in Mid-Miocene times. Folding and imbrication are directed to the northeast towards the recent foredeep of the Po Plain and the Adriatic Sea. The whole orogen lies on the western rim of the Adriatic micro-plate the basis of which sinks down to almost 40 km under the Po Plain (cf. Fig. 20).

Coalification studies have been performed by REUTTER et al. (1978,1982,1983) on the basis of 180 samples collected mainly from outcrops of molasse and flysch sediments.

As in the foredeep of the northern Alps, in the Apenninic foredeep the degree of coalification rises only slowly with depths as is shown in Fig. 18 for the 5320 m deep borehole Ponte dell'Olio situated 20 km south of Piacenza (cf. Fig. 19). In this borehole the vitrinite reflectance increases from 0,4 to 0,6 % Rm over a vertical distance of 3000 m in the autochthonous Miocene, belonging to the folded sub-Apenninic sediments of the Po Plain. The Miocene is overlain by a low rank Palaeocene-Eocene flysch nappe.

The coalification gradient in the Miocene is 0,07 % Rm/km, the geothermal gradient is 19° C/km.

A much higher coalification gradient was found on the southwestern side of the Apennine Mts., in the backdeep, where a gradient of 0,19 % Rm/km was determined in the Martina 1 borehole, situated offshore 27 km south of the southern coast of Elba

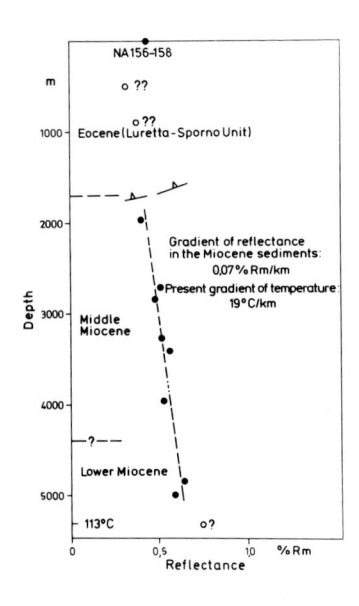

Fig. 18. Increase of vitrinite reflectance with depth in the Ponte dell'Olio borehole. Coalification is slightly more advanced in the Liguride-Luretta-Sporno nappe than in the underlying autochthonous sediments of the Po Plain. Note the very low reflectance values at a depth of 5000 m. They reflect the very reduced heat flow of the sub-Apenninic part of the Po Plain (after REUTTER et al., 1983).

Island in the Tyrrhenian Sea. There, the present geothermal gradient is 32° C/km (REUTTER et al., 1983).

Fig. 19, a coalification map of the northern Apennines, reveals the increase of rank from the external part of the mountain chain, bordering the Po Plain, to the internal zone near the Ligurian and Tyrrhenian coast. This trend is valid not only for all samples measured, but also for samples belonging to a single tectonic unit. This suggests that heating was stronger in the internal zone of the orogen than in the external zone. The isolines of present heat flow density in Fig. 19 strike from northwest to southeast, the values increasing from 30 mW/m² in the northeast, near the Po Plain, to more than 110 mW/m² in the southwest at the Tyrrhenian coast. The anomalously high heat flow in the southwest corresponds to the very high rank of coal in the same region. Near Grosseto, coals of uppermost Miocene age reached the rank of bituminous coal (0,74 % Rm at Ribolla), whereas coals of the same age are still in the stage of soft brown coal at the border of the Po Plain (0,22 % Rm). The strong coalification in the backdeep of the orogen, i.e. at the Tyrrhenian coast, is caused by very young (post-orogenic) granite intrusions. In the large nappe anticline of the Apuan Alps, east of Carrara, vitrinite plant remains embedded in Oligocene and Miocene sandstones, are even converted to graphite with reflectance values of more than 16 % Rmax. Here, the metamorphism occurred not earlier than 18 million years ago.

In the northern Apennines, according to REUTTER et al. (1983) the main coalification was late-synorogenic. It took place during the latest Miocene when the crust had returned more or less to the present level. Thus, if the younger contact metamorphism is disregarded, regional heating cannot have lasted more than 16 m.y., and as heat flow increased continuously starting from very low values, only the last part of this time span was decisive for the present degree of coalification. This conclusion of REUTTER et al. (1983) explains the close relationship between the coalification pattern of the northern Apennines and the present heat flow as well as with the present depth of the crust/mantle boundary. These relationships are demonstrated in Fig. 20, a cross-section through the northern Apennines. It shows the subduction of the Adriatic crust as well as the intensive imbrication of nappes below the Po Plain. There, the thickness of Pliocene + Pleistocene reaches more than 5000 - 6000 m and the Mohorovičić discontinuity dips down to 38 km near Faenza at the southeastern margin of the Po Plain.

The Atreo 1 borehole reached the Pliocene basis at a depth of 4997 m (AGIP 1977) whereas the Nonantola 1 borehole had not reached the basis of the Pliocene at 5809 m (AGIP 1977). The anomalous great thickness of the Pliocene/Pleistocene in the same region under which the Moho sinks to great depth suggests a recent subsidence of the crust/mantle boundary. If this is so, the subsidence rate of the upper mantle surface would be 1 mm/year, suggesting a great mobility of the Mohorovičić discontinuity.

As shown in Fig. 20 the Moho ascends from the Po Plain towards the southwest under the Apennine Mountains, up to 20 km under Elba Island. Correspondingly, the heat flow rises from 40 mW/m² in the Po Plain to 110 mW/m² at the Tyrrhenian coast. An anomalously high maximum of heat flow density (320 mW/m²) is reached near Larderello where geothermal energy is used economically. In accordance with the present heat flow and with the present depth of the Mohorovičić discontinuity, the degree of coalification is low in the northeast (0,5 % Rm) and rises towards the southwest to 1,4 - 1,6 % Rm in the Pratomagno Mts. and to 4,9 - 5,5 % Rm in the Colline Metallifere and on Elba Island (Fig. 20).

Moreover, the section of Fig. 20 demonstrates the different stages of advancing orogeny from southwest to northeast: downgoing crust under the Po Plain, compressional tectonics in the external side of the Apennines, uprising (detached) crust and tensional tectonics in the internal zone (southern Tuscany). The anomalous crust-mantle transition in the internal zone is obviously related to the high heat flow and the magmatic activity. In summary, Fig. 20 demonstrates the contrast between subsidence, compression, thickening of the crust and weak coalification in the foredeep, and uplift, tension, plutonism and strong coalification in the backdeep of the northern Apennines.

74

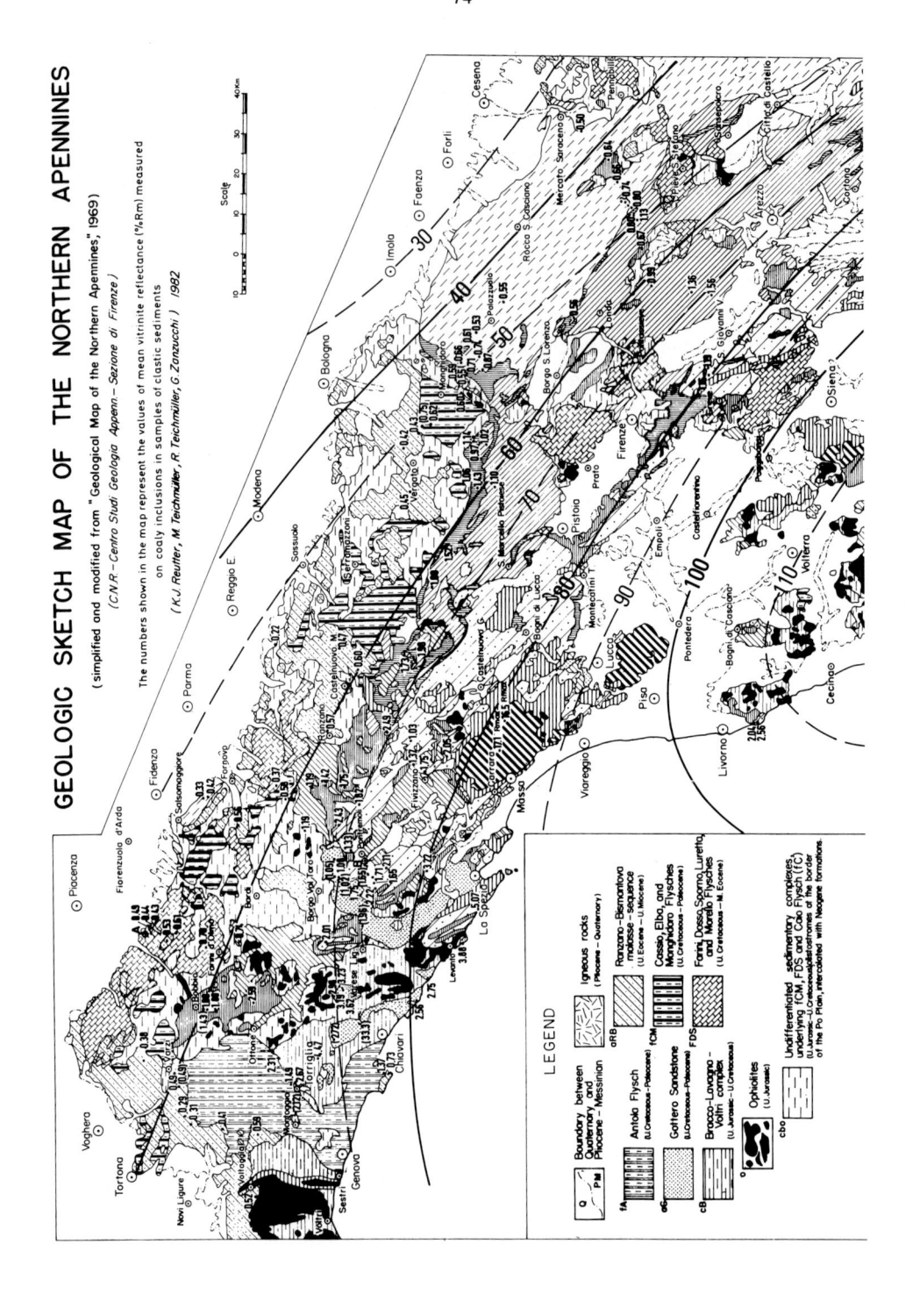

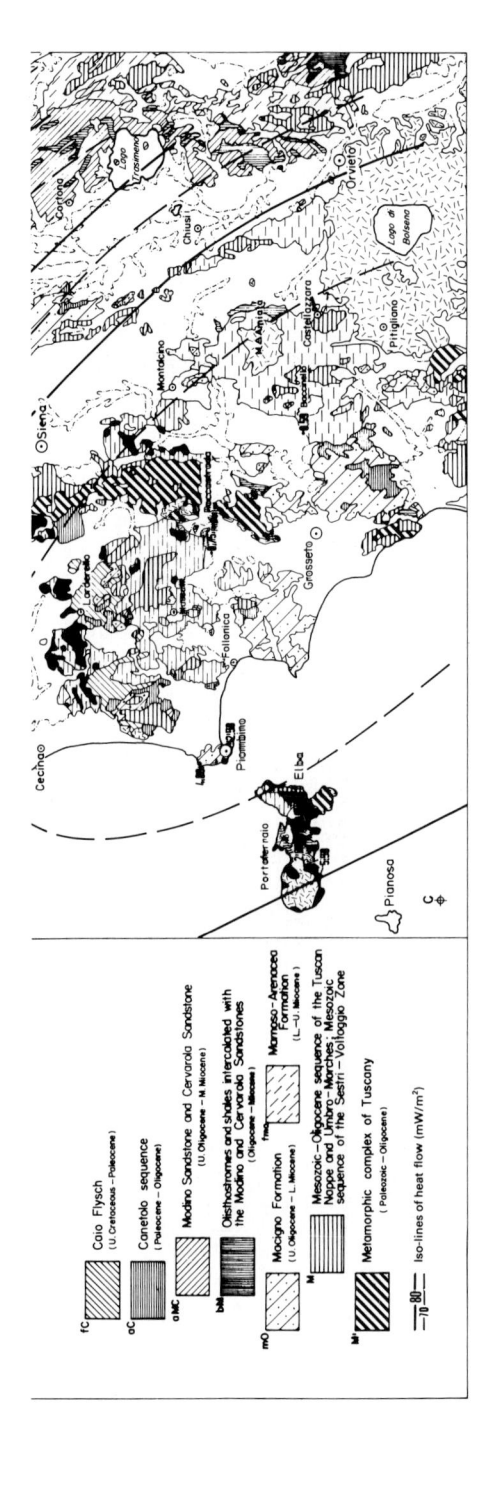

Fig. 19. Coalification map of the northern Apennines (after REUTTER et al., 1983) with iso-lines of heat flow density (after ČERMAK & HURTIG, 1979)

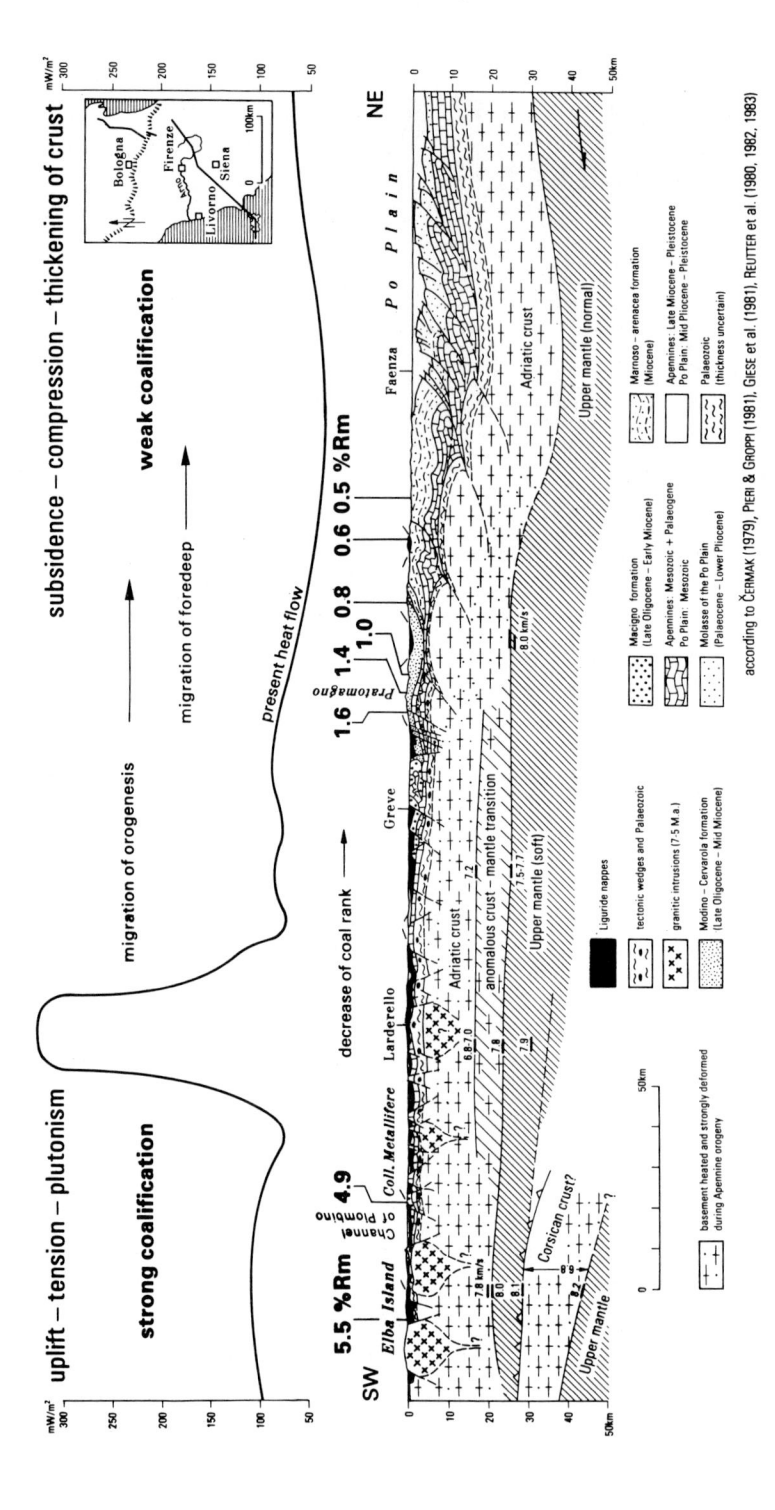

Fig. 20. Crustal cross-section through the northern Apennines between Elba Island and the Po Plain with values of vitrinite reflectance (% Rm) and of present heat flow density (mW/m²) (after REUTTER et al., 1983)

Conclusions

The marked difference between coalification and geothermal conditions in the sub-Variscan foredeep of the Ruhr-Lower Rhine Basin and the sub-Alpidic foredeeps of the Alps and northern Apennines might be explained by differences of crustal thickness. According to ČERMÁK (1979) and ČERMÁK & ZAHRADNIK (1982) it seems justified that the high geothermal gradients, deduced from the high coalification gradients of the Ruhr Carboniferous, point to a thin crust during the Westphalian. On the basis of heat flow density, deduced from an estimated mean thermal conductivity and the calculated palaeogeothermal gradients, KOPPE (1980: 61) estimated a crustal thickness of 19 - 23 km for the Ruhr-Lower Rhine area in Westphalian times, facing a thickness of 29 km at the present time (PRODEHL et al., 1976). Fig. 21 shows the inverse relation between heat flow density and crustal thickness for the Bohemian Mass, the Pannonian Basin and the Upper Rhine Graben according to ČERMÁK (1979), together with the values of KOPPE (1980) for the Westphalian A and the Westphalian B/C at the Ruhr. We introduced data for the Variscan backdeep of the Saar-Nahe Basin obtained by BUNTEBARTH (1983) as well as values reported in this paper from the foredeep molasses of the northern Alps (JACOB et al., 1982) and the northern Apennines (REUTTER et al., 1983) and from the backdeep of the northern Apennines (REUTTER et al., 1983). The diagram suggests:

1) a much thinner crust in the Subvariscan foredeep during the Upper Carboniferous (19 - 23 km) than in the Alpidic foredeeps of the northern Alps and the northern Apennines (33 - 41 km),

2) higher heat flow densities and thinner crusts in the backdeeps of the Nahe Basin (Permian) and the northern Apennines, compared with the corresponding foredeeps.

The high heat flow and the thin crust estimated for the sub-Variscan foredeep correspond to the high geothermal gradients estimated by ZWART (1967, 1976) for the European Variscicum. ZWART deduced the high geothermal gradients from the low thickness of the rock metamorphic zones ("low pressure metamorphism") and the unusually large number of granites and granodiorites which intruded during the time span between 350 and 280 million years before the present. He supposes that activity in the mantle, possibly as mantle diapirs, are the primary cause of the extrusive thermal activity in the last half of the Palaeozoic. ZWART (1976) writes: "As indicated by the low pressure mineral assemblages metamorphism took place at relatively shallow depths varying probably between 5 and 15 km. Therefore geothermal gradients must have been quite high generally about 40 to 60° C/km, but locally much higher up to 100° C/km or even more."

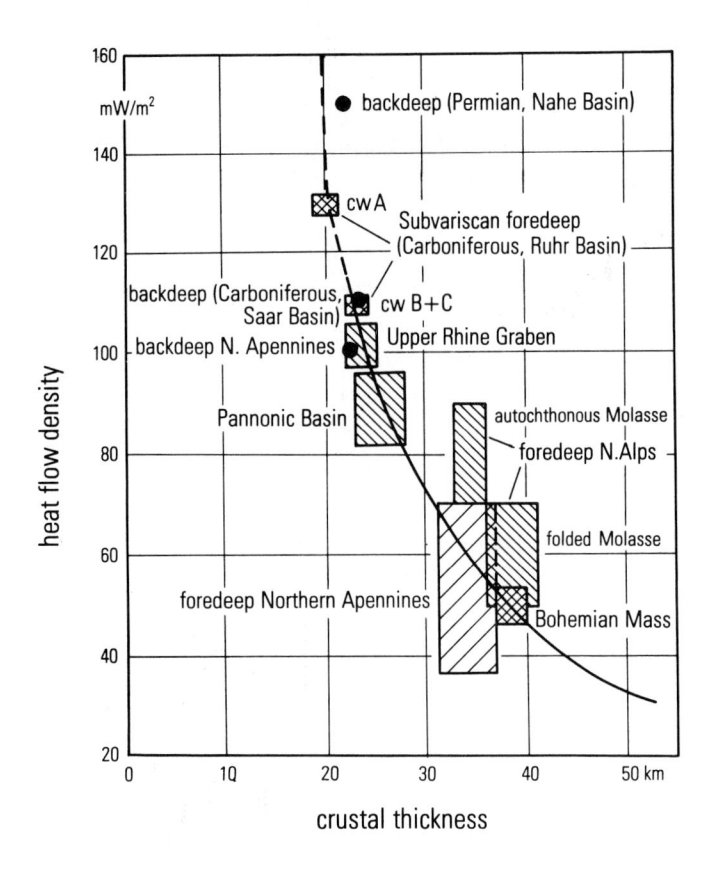

Fig. 21. Relationship between heat flow density and thickness of the crust (after ČERMÁK, 1979; KOPPE, 1980).

The high palaeogeothermal gradients in the sub-Variscan foredeep correspond also to the age of rock metamorphism in the northeastern Rhenish Variscan Mountains where, according to results obtained by the K/Ar-method (AHRENDT et al., 1976), the peak of metamorphism occurred at about 300 million years before the present, i.e. in the Westphalian B (WEBER & BEHR, 1983). According to BUNTEBARTH et al. (1982) this coincidence emphasizes a heat supply from greater depth effective in both regions, the northeastern Rhenish Variscan Mountains and the sub-Variscan foredeep of the Ruhr-Lower Rhine Basin.

The surprisingly low coalification gradients and low geothermal gradients in the foredeeps of the northern Alps and northern Apennines correspond with the thickening of the crust in these regions. In both cases the great crustal thickness is caused by subduction of cold continental plates. Conversely one may come to the conclusion that subduction did not take place in the sub-Variscan foredeep of the Ruhr-Lower Rhine area.

ADDENDUM

We now think it is better to use a regression method that allows error to be present in both the independent and dependent variable. The major axis regression method calculates the following equation from the same set of R_m and temperature data: $\ln(R_m) = 0.0096 \,(T_{max}) - 1.4$

This equation produces peak temperature predictions from R_m that are more compatible with other geothermometers and thermal history reconstruction. This equation should replace the one given in the abstract on page 79, and in the text on page 83.

Lecture Notes in Earth Sciences, Vol. 5
Paleogeothermics. Edited by G. Buntebarth and L. Stegena.
ISBN 3-540-16645-9
© Springer-Verlag Berlin Heidelberg 1986

THE CORRELATION OF VITRINITE REFLECTANCE WITH MAXIMUM TEMPERATURE IN HUMIC ORGANIC MATTER

BARKER, Ch.E. and M.J. PAWLEWICZ
U.S. Geological Survey
Box 25046, MS 921, Denver, Colorado 80225, USA

Abstract

Mean random vitrinite reflectance (Rm in %) shows a strong correlation ($r^2 = 0.7$, n > 600) with maximum burial temperature (T_{max} in °C). These data are modelled by the linear regression equation:

$$\ln(Rm) = 0.078\ T_{max} - 1.2$$

T_{max} and Rm were compiled from over 35 systems, rich in humic organic matter to minimize the effect of chemical composition on Rm. The thermal maturation data span a range from early diagenesis to greenschist metamorphism over a T_{max} interval from about 25 - 325° C and 0.2 - 4.0 % Rm. Burial history reconstruction indicates that the functional heating duration (elapsed time as temperature increases within 15° C of T_{max}) of these systems ranges from 10,000 yr to more than 10 m.y. The strong correlation of T_{max} with Rm, irrespective of functional heating duration and in diverse geologic systems, indicates that increasing time at T_{max} has little influence on thermal maturation of sedimentary organic matter. Instead, uncertainty in correction of borehole temperature logs, T_{max} determination, and Rm measurement explains much of the remaining variability not accounted for by the regression analysis. We did not attempt to correct the measured borehole temperature to equilibrium reservoir conditions because there is no consensus on which method to use, the necessary data is often unrecorded, and predictions made from our T_{max} - Rm calibration are compared to uncorrected T_{max} data. We found that T_{max} is difficult to determine in sedimentary environments that have cooled because of the weak thermal imprint on the rocks in low temperature systems and poorly-known burial histories. Variability in Rm measurement appears mainly due to operator or laboratory bias, increasing bireflectance with rank, and variation in diagenetic history which causes reflectance suppression.

These studies imply that T_{max} controls Rm, making the relationship useful as a maximum geothermometer, but that several physico-chemical and technical factors obscure the correlation. The problems in measuring T_{max} and Rm, shown by the appreciable data scatter, make our calibration imprecise. However, application of this geothermometer to systems where T_{max} is well known shows that it yields realistic paleotemperature estimates. Other support for temperature control of Rm is documented from studies of metamorphic mineral assemblages and coal rank, and critical testing of temperature-time-rank models in sedimentary systems.

Introduction

Maximum burial temperature and heating duration (or geologic time) are commonly reported as the most influential controls in the thermal maturation of sedimentary organic matter (OM) (WAPLES, 1984). Recent studies suggest that the effect of heating duration is limited, and relatively soon (in a geologic time frame) after reaching maximum temperature, OM stabilizes and ceases significant reaction (BARKER, 1983; PRICE, 1983; and others). Under these conditions, mean random vitrinite reflectance (Rm), a measure of thermal maturation, should be a function of the maximum temperature reached in the system. The question of whether heating duration is of continuing importance in thermal maturation of OM can be addressed by plotting a scatter diagram of the maximum burial temperature (T_{max}) versus Rm for samples from sedimentary systems with a wide range of heating duration. A near-zero coefficient of correlation (r^2) calculated for these data would indicate that these data vary randomly with respect to each other and that a third variable (presumably heating duration) has a significant effect on the system. Conversely, a r^2-value approaching one would indicate that T_{max} and Rm are strongly dependent, and that heating duration cannot be a significant factor in the thermal maturation of OM over extended geologic times.

Our approach to calibrating a maximum-recording geothermometer based on thermal maturation of OM is to compile published T_{max} and Rm data from systems with a wide range of functional heating duration. Regression analysis of T_{max} and Rm indicates whether this system can be adequately characterized by considering only these two variables. This approach is necessary because of the uncertainty in sedimentary systems in determining a chemically meaningful heating duration (an estimate of reaction time), or a temperature of chemical reaction. Coupled with only a schematic understanding of the reaction mechanisms in OM (TISSOT & ESPITALIÉ, 1975) it is difficult to define rigorously thermal maturation as a chemical system. The advantage of this approach is that the data is compiled from numerous independent measurements in diverse sedimentary systems. The limitations of our empirical method to modelling OM thermal maturation are that: (1) maximum temperature reached in the system may be difficult to determine; (2) correction of borehole temperature data to equilibrium reservoir conditions is poorly understood; and (3) Rm data may be subject to significant error. Our model is a simple approximation of a complex chemical system. However, as shown by numerous studies (WAPLES, 1984), and by examples presented in this paper, OM thermal maturation is successfully modeled by simple empirical functions.

Data and Results
Maximum Temperature and Vitrinite Reflectance

T_{max} and Rm data were compiled from over 35 systems (boreholes in sedimentary basins) undergoing conditions ranging from burial diagenesis to greenschist metamorphism. Only rich in humic OM were used because of the influence of chemical compositions on Rm. Maximum vitrinite reflectance data was converted to mean random vitrinite reflectance using the equation of TING (1978). The geothermal curve was determined by computing a geothermal gradient from bottom hole temperature (BHT) and the mean annual surface temperature. The reported sample depth was used to compute the system temperature at that point by interpolation from the geothermal curve. The T_{max} condition in the system was indicated by geologic and stratigraphic reconstruction (WAPLES, 1981) or comparative geothermometry (BARKER & ELDERS, 1981). The variability in approach to determining T_{max} and Rm makes this a heterogeneous data set. We generally had to accept both the T_{max} and Rm data without confirming their accuracy. Only systems where apparent maximum temperature existed or could be interpreted from other evidence are used to calibrate this OM thermal maturation model.

Heating Duration

Methods of computing heating duration for OM thermal maturation are attempts to estimate reaction time during burial where temperature slowly changes, or stabilizes, over an extended geologic time. HOOD et al. (1975) defined the effective heating-time for OM thermal maturation as the elapsed time when the system is within 15° C of maximum temperature. MACKENZIE & McKENZIE (1983) have demonstrated that this is a reasonable measure of reaction time in sedimentary systems. We agree with their conclusion except that thermal maturation of OM, given sufficient time, could stabilize by reactions consuming all potentially cleavable bonds at maximum temperature. Experiments indicate that thermal maturation of OM proceeds by parallel reactions that have a wide range of activation energy (E_a) (JÜNTGEN & KLEIN, 1975; TISSOT & ESPITALIÉ, 1975). For a reasonable burial temperature, the wide E_a range over which OM thermal maturation occurs suggests that the reactions: (1) with a low E_a are complete and not generating products; (2) with a moderate E_a are generating significant products; or (3) with a high E_a will be slow and not complete reaction in geologic time. The general correlation of Rm with T_{max} (NERUCHEV & PARPAROVA, 1972; SUGGATE, 1982; and others) suggests that only a limited suite of reactions control the thermal maturation of OM, and additional heating duration will not make the slower (high E_a) reactions significant in increasing Rm. Thus, a plausible heating duration estimate would be the elapsed time at maximum temperature necessary for the controlling reactions to approach completion. Thermal maturation of OM would stabilize at this point with respect to this reaction temperature, and heating duration is no longer a factor. We do not imply that thermodynamic equilibrium is established in the system because OM thermal maturation reactions are irreversible (BLUMER, 1965; TISSOT & WELTE, 1984). The irreversible nature of thermal maturation reactions cause the rank of OM, and consequently Rm, to be set by the maximum temperature.

Estimates of heating duration can emphasize elapsed time near T_{max} because lower temperature reactions are not significant in determining final OM rank. For example, OM thermal maturation, using the classic approximation that the reaction rate doubles for each 10° C increase, will be approximately 1000 times faster at 150° C (approximate cessation of liquid hydrocarbon generation) than the same reaction at 50° C (approximate initiation of hydrocarbon generation) (HUNT, 1979). The contribution to OM rank from low temperature reactions appears to be overwhelmed by reaction at higher temperature.

These considerations indicate that the effective heating-time of HOOD et al. (1975) should be more limited and not consider time elapsed during declining or stable temperature. A functional heating duration for OM thermal maturation is defined as the elapsed time while temperature increases within 15° C of maximum temperature. Geologic time during temperature decline in the systems is not considered to increase Rm because OM thermal maturation is irreversible. Also, if the temperature stabilizes at near-maximum, this time is not considered to increase effectively the thermal maturation of OM and is not included in the functional heating duration.

The functional heating duration is still a contrived estimate of reaction time. There is no known method of determining a reaction time in sedimentary systems that is applicable to kinetic equations. The control of OM rank by irreversible reactions at T_{max} indicates that reaction time in sedimentary systems is the elapsed time at maximum temperature for the controlling reactions to approach completion. However, this type of heating duration estimate is not easily measured in sedimentary systems where temperature has increased slowly during burial (in this case, time at $T_{max} = 0$), or have poorly known burial histories, making this an impractical definition. The definition of the functional heating duration over a 15° C range near T_{max} is a reflection of imprecise geologic data.

Regression analysis

Proper designation of the dependent and independent variables for regression analysis of T_{max} - Rm data and calibration of an empirical model are crucial (BARKER, 1984). In sedimentary basins, sample temperature is determined by the geothermal curve and depth within the system. The independent variable (T_{max}) is fixed by the investigator by selecting a sample at some depth, and the dependent variable (Rm) is measured on OM concentrated from that sample. In this case, estimates of T_{max} from Rm must be made from the regression of dependent variable (Y) and independent variable (X) (SNEDECOR & COCHRAN, 1967). Regardless of the statistical theory and its constraints, the selected regression curve should be a good model of the data and effectively predict its trend. Inspection of the data set suggests an exponential trend [$Rm = a* \exp(b T_{max})$] or its linear transform $\ln(Rm) = a + b(T_{max})$ as used in this paper.

Vitrinite reflectance in these sedimentary basins is highly dependent on maximum temperature (Fig. 1). A least squares regression equation:

$$\ln(Rm) = 0.0078 T_{max} - 1.2$$

computed from over 600 maximum temperature (in °C) and vitrinite reflectance (in %) data from these systems, indicates that about 70 percent ($r^2 = 0.7$) of the variability in the vitrinite reflectance data is explained by considering temperature alone.

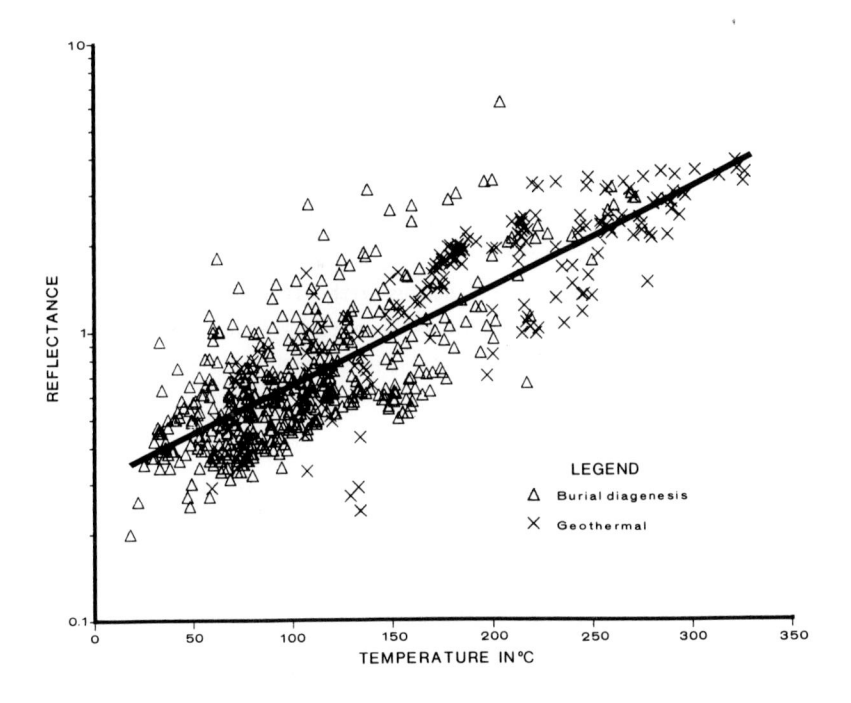

Fig. 1. General correlation of vitrinite reflectance with maximum temperature. Least squares regression analysis indicates that the variables are strongly related with correlation coefficient (r^2) = 0.7 for a sample size (n) > 600. The functional heating duration varies from 10^4 to 10^8 years in these systems. Sources of data used in this figure are available upon request.

Heating Duration and Thermal Maturation

The influence of heating duration is assessed using the LOPATIN (1971) model, which generates an estimate of OM thermal maturity by numerical integration of heating duration in each 10° C interval over the system's burial history. The Lopatin model is modified in our analysis because the functional heating duration is computed using only the elapsed time in a small temperature interval near T_{max}, making numerical integration of time and temperature unnecessary. OM thermal maturity then equals elapsed-time multiplied by $2^{T_{max}/10}$. If temperature and heating duration both determine OM thermal maturation, then the range of T_{max} across a line of constant Rm would be due to heating duration. For OM thermal maturity to remain constand requires that the T_{max} factor changes in an opposite manner to the heating duration factor. T_{max} ranges over about 100° C across a given isoreflectance line in our data (Fig. 1) or over a factor of 1,000 in the Lopatin model. The functional heating duration would also range over a factor of 1,000. Using a functional heating duration of about 1 m.y. for burial diagenesis indicates a range up to 1 b.y. to compensate for the temperature factor. This is an unreasonably large effect for heating duration because geologic studies show that OM rank is only influenced by temperature after about 10 m.y.

Geologic studies indicate that OM thermal maturation does stabilize after about 10^6 - 10^7 year (Table 1) and in our model increased functional heating duration after stabilization would produce negligble increase in OM rank. Burial history reconstruction of the systems we studied, and others with different OM types, shows that about 90 percent have been within 15° C of maximum temperature for greater than 10^6 yr (Fig. 2). Thus, in most cases of burial diagenesis, heating duration at maxi-

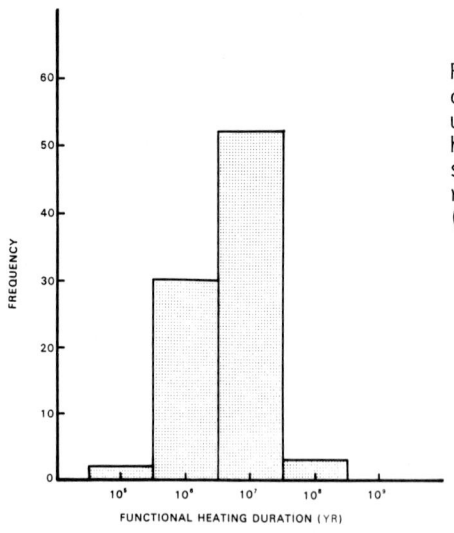

Fig. 2. Histogram of functional heating duration in selected sedimentary basins undergoing burial diagenesis. The functional heating duration in the nine geothermal systems used in Fig. 1 (not plotted here) range from 10^4 - 10^6 years. Data from BARKER (in press).

Table 1. Published estimates of the time required for the stabilization of OM thermal maturation. Insignificant stabilization time indicates that OM thermal maturation was found to be controlled by temperature.

Reference	Stabilization Time (yr)	Notes
Seyer (1933)	short	Petroleum generation above 200° C
McNab et al. (1952)	10^6	Petroleum generation.
Vallentiyne (1964)	10^6	Complete amino acid decarboxylation at 100° C.
Tan (1965)	$6 \cdot 10^7$	Coal maturation.
Abelson (1967)	10^6	Methane generation from oil shale pyrolysis would be complete in 10^6 yr at 115° C.
Abelson (1967	10^6	Analysis of Los Angeles hydrocarbon generation data (Phillipi, 1965) suggests effective duration of the total (heating) exposure is equivalent to roughly $2 \cdot 10^6$ yr at 150° C.
Brooks (1970)	Insignificant (Waples,1984)	States that coal rank is a maximum-recording geothermometer.
Neruchev & Parparova (1972)	10^6	Coal and kerogen maturation.
Lopatin & Bostick (1973)	Insignificant (Waples,1984)	State that coal rank is a maximum-recording geothermometer.
Bartenstein & Teichmüller (1974)	Insignifcant (Waples,1984)	State that col rank is a maximum-recording geothermometer.
Nagornyi & Nagornyi (1974)	Insignificant	"Geological time does not limit the coalification process."
Juntgen & Klein (1975)	10^7	Coal maturation.
Hacquebart (1975)	10^7 to 10^8	Coal maturation.
Ammosov et al. (1975,1977)	10^6	Heating time not important in organic maturation.
Demaison (1975)	10^8	Coal maturation.
Harwood (1977)	Short	Kerogen pyrolysis.
Cornelius (1975)	about 10^7	Equilibrium reached in petroleum generation.
McTavish (1978)	Insignificant	The time factor for petroleum generation is of limited significance over geologic time and probably critical only for a short period.
Veto (1980)	finite	"A minimum temperature is needed to start any transformation of organic material with a particular activation energy. At such a temperature a certain length of time is necessary and sufficient to comple the reaction .."
Veto (1980)	Time effect exaggerated	"The role of time is probably also exaggerated .. for the methods of Bostick and Lopatin."
Barker, Colin (1979)	about $5 \cdot 10^7$	Influence of time is minor when compared to the effect of temperature in hydrocarbon generation from kerogen.
Sajgo (1979)	10^6	Petroleum generation from kerogen.
Wright (1980)	Insignificant	"Temperature remains critical: a source shale with $R_0=0.8\%$ can remain at that rank for many million years and never generate a drop of oil."
Teichmüller & Teichmüller (1981)	$3 \cdot 10^6$	Equilibrium in organic maturation not established in this heating time.
Price et al. (1981)	Short	Organic metamorphism in hydrothermal bombs at 350° C.
Suggate (1982)	10^6	Coal maturation.
Gretener & Curtis (1982)	Insignificant	Petroleum generation above 130° C.
Barker (1983)	10^4	Kerogen maturation in liquid-dominated geothermal systems.
Price (1983)	10^6	Kerogen maturation.

mum temperature has been sufficient for the controlling reactions to approach completion and for the OM to stabilize effectively with respect to temperature. Heating duration would have no further effect.

Our interpretation is that the so-called "influence of time" is invoked to explain differences in temperature-Rm curves between comparable sedimentary systems - some now at T_{max} and others that have cooled. For instance, the Munsterland-1 well used by LOPATIN (1971) to calibrate his OM thermal maturation model has declined significantly from maximum temperature (Fig. 3) (PRICE, 1983). Temperature decreases causes Rm data to be shifted to the left from their original position at T_{max} (Fig. 3). This shift is similar to the effect increased heating duration would have on OM thermal maturation.

The remaining variation in the T_{max} and Rm data could be accounted for by considering functional heating duration but other significant types of data variability minimize the contribution from this source.

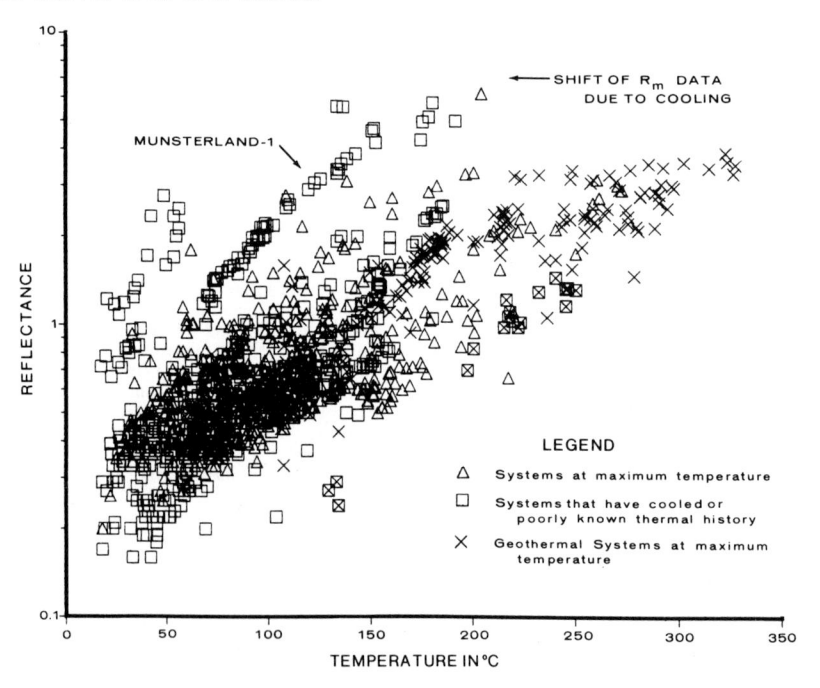

Fig. 3. Scatter plot of Rm and temperature data including those not at maximum temperature. Rm and temperature data from systems in which temperature has declined are shifted to the left from the systems now at T_{max} shown in Figure 3. The Munsterland-1 well shown was used by LOPATIN (1971) to calibrate his time-temperature index OM maturation model. Sources of data used in this figure are available upon request.

Errors in Vitrinite Reflectance and Temperature Data
Vitrinite Reflectance

Operator bias in selecting vitrinite for measurement is potentially the greatest source of error during reflectance analysis. Bias can occur in reflectance measurements when multiple vitrinite populations are present in the sample requiring the operator to select a single population for analysis. These mixed populations arise primarily from the admixture of recycled, previously-altered vitrinite with first cycle OM during sedimentation. To minimize this problem and compile a consistant data set, one microscopist should make the Rm determinations (although a single operator could still consistently shift the Rm measurements). However, because our data was in part derived from the published work of different microscopists, variable operator-bias is a significant factor in causing data scatter. Rm measurements by thirty microscopists on the splits of 19 different samples show that the range of Rm measurement can be up to +/- 0.4 % Rm in low rank OM (unpublished report, International Commission on Coal Petrology, see BOSTICK, 1979). This wide Rm range arises from operator bias and differences between laboratories in processing, polishing, and photometer calibration. Even under an optimal scenario to minimize operator and laboratory bias - microscopists measuring the same uniformly prepared sample suite, using the same microphotometric system, and similar vitrinite selection procedures - Rm determinations still vary by up to -/+ 0.2 % at moderate to high rank (BARKER, 1983).

Mixing of drill cuttings in the borehole can also introduce multiple vitrinite populations in a sample. This is a major physical problem in vitrinite reflectance analysis because drill cuttings are the only rocks available over a significant depth interval in most boreholes. Up-hole sloughing of rock from the sides of the borehole introduces a less mature vitrinite population, that if abundant, could be mistaken for the first-cycle (lowest maturity) population indigenous to that sample. Up-hole sloughing of rock can reduce the apparent Rm to the level of the last casing depth, which in the worst case would be that of the near-surface rock. Rock-chip mixing in the drilling fluid and infrequent sample collection make drill cuttings a composite sample (at best) of the drilled interval between sample points even when the sample depth is corrected for transit time from drill-bit to surface. This introduces uncertainty in where to determine the sample depth, thus causing error in T_{max} and Rm data. It also appears possible to severely alter the rock chips by heat and pressure from the drill bit (TAYLOR, 1983), though these are probably identifiable in a drill-cuttings sample. The less altered rock, however, may contain an artificially matured OM, without visibly changing the petrographic character, making them undetectable.

Another significant source of error in vitrinite reflectance analysis is Rm suppression. Studies have shown that Rm can be suppressed up to several tenths of a percent by maceral association and differences in early diagenetic history (see review by PRICE & BARKER, 1985). Variation in the depositional and (or) diagenetic environment produces a hydrogen-rich OM that has lower Rm values than those expected from the thermal history of the sediment. This problem was minimized in this study by considering only those systems rich in humic (type III) OM.

Vitrinite also becomes anisotropic or bireflectant of moderate to high rank. The physical effect is that reflectance becomes dependent on the microscope stage orientation. Bireflectance is insignificant below about 1.2 % Rm, but the difference between the maximum and minimum reflectance increases to almost 25 % at 3.0 % Rm (STACH et al., 1982). Bireflectance in high rank OM increases the range of the reflectance histogram and often produces polymodal distributions, making the mean value less representative. Further, there is a convergence of the reflectance for vitrinite and inertinite at about 2.5 % Rm which makes these two macerals difficult to distinguish.

Temperature

Most T_{max} data for systems now at peak temperature are from uncorrected BHTs typically taken on a single logging tool run. A minority of the remaining temperature data are from corrected BHT measurements (the correction method sometimes unspecified). A geothermal gradient is calculated from BHT by interpolating between mean annual surface temperature and BHT. This linear approximation can be highly inaccurate because the temperature profile can change with lithology (thermal conductivity), subsurface fluid flow, etc.(DRURY et al., 1984). Accurate determination of present-day formation temperature also requires thermal equilibration of the borehole before measurements. The extended borehole shut-in time required to establish thermal equilibrium usually means that borehole temperature is measured soon after drilling is completed. Drilling disrupts the undisturbed temperature profile because cool fluids are pumped down the borehole and are warmed up on ascent to the surface. The deep portions of the borehole are cooled and the shallow portions are heated during drilling. The shallow portions of the borehole are exposed to more drilling fluid and cooling there is greater than at depth. BHT data are not usually confirmed as equilibrium reservoir temperatures by repeated measurements over a significant time interval making correction necessary. Our experience is that the data necessary for calculating a correction of the log temperature to equilibrium formation temperature are not included in the borehole history reports. Further, attempts at correcting the BHT measurement to equilibrium conditions, although necessary, are often unsuccessful because drilling and measurement conditions vary widely and usually cannot be corrected by some uniform procedure (DRURY, 1984). Temperature corrections in the order of 20 - 30° C are typical for BHT data (HOOD et al., 1975) making this error at least as significant as that from determining T_{max}.

T_{max} is a difficult determination to make accurate, especially in systems that have cooled. The burial history reconstruction method (see discussion by WAPLES, 1981) is widely used to estimate maximum burial depth (and temperature) during diagenesis. Temperature as a function of geologic time is calculated by using the existing or paleo-geothermal gradient applied to the depth-time curve constructed from the burial history. Lacking geologic evidence to detail thermal changes in the system, the calculated geothermal gradient is usually assumed to remain constant through time without regard to heat-flow change, diagenesis, and lithology. Thus, the detailed geologic analysis is reduced to time-temperature data by simplification of the changes that can occur in the geothermal gradient. Unfortunately, the geologic record often will not permit a more sophisticated approach and a rigorous definition of the time-temperature history.

T_{max} in systems that have cooled may also be determined by using maximum geothermometers. The limitation of this method is that the T_{max} event must be recorded in the rocks. Thermal events can be difficult to record in sedimentary systems because of the slowness of equilibration reactions at low temperature. The short-term nature of some thermal events may also preclude them from leaving an imprint on the rock. Both of these conditions produce significant changes in the OM rank (STACH et al., 1982).

Discussion

Three lines of evidence support our regression analysis indicating that maximum temperature alone determines Rm and heating duration has little continuing influence on OM thermal maturation: (1) tests of temperature-time-rank models in hypothetical or well-known sedimentary systems; (2) comparison of OM rank to mineral assemblages resulting from equilibrium reactions during metamorphism; and (3) application of this empirical T_{max} - Rm geothermometer to sedimentary systems where temperature is well known.

Tests of Temperature-Time-Rank Models

VETÖ (1980) tested published temperature-time models of OM thermal maturation in 45 sedimentary basins and found "the role of time is probably also exaggerated ... for the methods of BOSTICK and LOPATIN." He considered that "a minimum temperature is needed to start any transformation or organic material with a particular activation energy. At such a temperature a certain length of time is necessary and sufficient to complete the reaction...". WRIGHT (1980) in a similar test using hypothetical burial histories concluded that "temperature remains critical: a source shale with

R_o = 0.8 % can remain at that rank for many millions of years and never generate a drop of oil." Several other studies show that heating duration is not of continuing importance in OM thermal maturation (Table 1).

Temperature-time models are poor predictors of heating duration in geothermal systems when compared to direct estimates of thermal event duration. BARKER (1979) found the correlation (r^2 = 0.8) between Rm and logged temperature in the central portion of the Cerro Prieto system, together with consistent temperature estimates from fluid inclusion and oxygen isotope geothermometry, indicates that these rocks are now at maximum temperature. Application of this data to KARWEIL's (1956) OM thermal maturation model predicts a heating duration of 5 m.y. for the Cerro Prieto system - a poor fit to a heating duration of about 10,000 years indicated by fission-track annealing studies (SANFORD, 1981), and the upper limit of heating duration implied by reservoir rock age of about 1 m.y.

Metamorphic Mineral Assemblages and Rank

Stability of mineral assemblages resulting from hydrothermal metamorphism are temperature dependent. The consistent occurrence of authigenic minerals at similar temperatures in many rock types of different ages and different geothermal systems inditees that phase changes result from equilibrium reactions due to thermal metamorphism (ZEN & THOMPSON, 1974; BROWNE, 1978; WEAVER, 1979). The formation of equilibrium assemblages suggests that the reaction kinetics for these reactions are relatively rapid and that heating duration should not be a significant factor in hydrothermal metamorphism. BARKER et al. (in press) found that certain Rm levels correspond with hydrothermal zones in several boreholes in the Cerro Prieto geothermal system. These mineral zones occur over a wide temperature range, but the range of the thermal stability field and Rm is similar between boreholes. This relationship indicates that temperature controls Rm in hydrothermal systems.

KISCH (1969; updated by ZEN & THOMPSON, 1974) in a review of the available literature, found mineral zones associated with burial metamorphism correlated with coal rank in samples from various geographic localities and geologic times. Tertiary sediments in many areas of the United States show clear correlation between pressure-temperature (P-T) dependent mineral metamorphism and kerogen metamorphism (VAN DE KAMP, 1976). Similarly, STALDER (1979) has shown that clay mineral and zeolite mineral assemblages correlate with OM rank in the Eocene-Oligocene Taveyannaz Sandstone (Europe) and its equivalents. SHIMOYAMA & IIJIMA (1976) showed that rank of Japanese Tertiary coals correlate well with zeolite zonation, and concluded, rank increase is essentially temperature dependent. LANDIS (1971) found that progressive graphitization is related to metamorphic grade as defined by mineral assemblages, and concluded that kerogen graphitization "... is primarily dependent upon metamorphic temperature; pressure and variation in starting material presumably consititute secondary controls." HOWER & DAVIS (1981) converted published coal rank and associated mineral assemblage data from various geologic times to consistent set of P-T, and vitrinite reflectance values. HOWER & DAVIS noted that considerable uncertainty was involved both in rank conversion, and in placing the rank data at a single point, whereas the metamorphic mineral assemblage delimits on the P-T field. However, their Rm data can be plotted along simple, contourable patterns on the P-T field, indicating that time is a negligible factor and can be ignored in modeling OM metamorphism.

In summary, metamorphic mineral assemblages resulting from temperature and P-T dependent equilibrium reactions, in diverse metasedimentary systems with a wide range of heating durations, correlate with a specific coal rank. This suggests that the controls on metamorphic mineral assemblages appear to control Rm. We interpret these data to indicate that Rm is primarily controlled by temperature because static pressure apparently has little influence on Rm (STACH et al., 1982).

Applied Geothermometry: Case Studies

POLLASTRO & BARKER (in press) computed paleotemperatures from Rm, illite/smectite ratios in mixed-layer clay, and fluid inclusion temperatures in the Wagon Wheel no. 1 borehole, Green River Basin, Wyoming (Fig. 4). Using our empirical calibration, they computed maximum temperature which is close to the temperature calculated from clay mineral data at about 190° C. Uncorrected borehole temperature at this depth is now about 130° C. The slope of the present day borehole temperature measurements and the slope of temperatures interpreted from Rm are about 25° C/km. Fluid inclusion homogenization temperatures in quartz-filled fractures formed by late Tertiary deformation are equal to present-day temperature. An Rm surface intercept of 0.33 % indicates about 1700 m of section has been removed (DOW, 1977). Decreasing the burial depth by 1700 m in a geothermal gradient of 25° C/km would decrease borehole temperatures by about 40° C, making it consistent with the other geothermometric data. These data were interpreted to indicate that temperature has decreased by about 40° C, by uplift and erosion rather than decline in geothermal gradient, before the fractures were

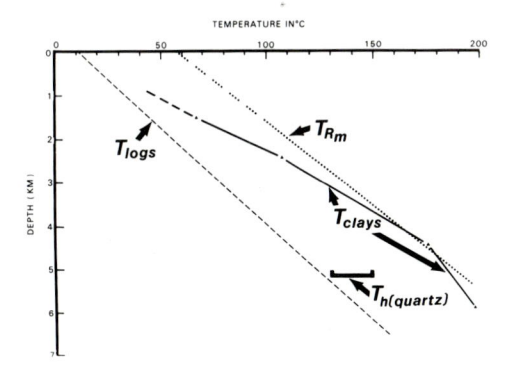

Fig. 4. Temperature versus depth, Wagon Wheel no. 1, northern Green River Basin, Wyoming. Temperature data from borehole logs (T_{logs}), clay minerals (T_{clays}), vitrinite reflectance geothermometry (T_{Rm}), and fluid inclusion homogenization ($T_{h-quartz}$). From POLLASTRO & BARKER (in press).

filled with quartz. A subsequent fission track annealing study in this same borehole confirmed a temperature decline of at least 20° C approximately 3 m.y. B. P. (NAESER, 1984).

BARKER & HALLEY (in press) studied paleotemperatures indicated by fluid inclusion, Rm, and oxygen isotope data from fracture-filling calcite and host rock in the Bone Spring Limestone, Permian Basin, Texas. Although the cements have petrographically distinct multiple zones, homogenization temperatures (T_h) of oil inclusions and $\delta^{18}O$ data divide these cements into two (early and late) groups. Early cements have $\delta^{18}O$ near 0 °/oo and contain fluid inclusions whose mean T_h is 70° C. The late cements have $\delta^{18}O$ of about -9 °/oo and contain sparse oil inclusions similar to those in the early cement. Primary fluid inclusions in the late cements have a mean T_h of 110° C. The mean T_h of these inclusions is close to the 120° C calculated from an Rm of 0.7 % with our empirical calibration and the 45° C temperature change from the early to late cement interpreted from $\delta^{18}O$ difference of about 9 °/oo. These thermal data and burial reconstruction suggest that both the late cements and Rm record the maximum temperature attained in the rock.

BARKER (1983) using empirical techniques similar to those in our study, compiled temperature, vitrinite reflectance, and heating duration data from six geothermal systems. Geothermometers indicate that present-day temperatures are close to the maximum temperature reached in these systems. These geothermometers were used to indicate: (1) ambient thermal conditions - by present borehole temperature, silica and sodium-calcium-potassium geothermometers, and the calcite-water oxygen isotope geothermometers; (2) short term thermal events - by fluid inclusion geothermometry on vein minerals; and (3) maximum temperature - from hydrothermal mineral zones. These geothermometers generally agree within a small range, suggesting that these systems are now at maximum temperature. Heating duration was determined by fission track annealing, and paleomagnetic and radiometric dating of the igneous heat source for the system. The concise definition of the maximum temperature and heating duration in these systems indicates that OM thermal maturation is dependent only on the maximum temperature reached after an elapsed time of about 10,000 years.

Summary

The evidence from temperature-time rank models, metamorphic mineral assemblages, case studies, and our regression analysis argue that functional heating duration has a limited influence on OM thermal maturation, and vitrinite reflectance can be directly calibrated as a maximum geothermometer. This relationship implies that because OM thermal maturation does record the peak temperature, the chemical reactions must stabilize. The limitations in calibrating Rm as a geothermometer are physico-chemical and technological problems in the measurement of OM thermal maturation and maximum temperature. The problems in determining Rm alone can explain most of the data scatter in Figure 1. These problems will not allow the definition of a concise relationship between T_{max} and Rm, and make temperature prediction from our model imprecise. This deficiency can be minimized by a careful approach to sampling and analysis by a single laboratory, illustrated by the excellent correlation ($r^2 = 0.9$) between T_{max} and Rm found by PRICE & BARKER (in PRICE, 1983) versus that found in this study (0.7). Application of this Rm-based geothermometer to systems where T_{max} is well known shows that it yields realistic paleotemperature estimates.

A COMPARISON OF TWO VITRINITE REFLECTANCE METHODS FOR ESTIMATING PALEOTEMPERATURE GRADIENTS

BUNTEBARTH, G.* and M. MIDDLETON**

* Institut für Geophysik, TU Clausthal
Arnold-Sommerfeld-Str. 1, D-3392 Clausthal-Zellerfeld, F.R. of Germany

** Geological Survey of W.A.
66 Adelaide Terrace, Perth 6000, Australia

Abstract

Two methods of determining the paleogeothermal gradient from vitrinite reflectance versus depth data are compared by application to 8 boreholes. One method, developed by BUNTEBARTH proposes that vitrinite reflectance is essentially proportional to depth (for constant burial rate and thermal gradient), whereas the second method, developed by MIDDLETON, proposes that vitrinite reflectance increases exponentially with depth. The two methods give similar paleothermal gradients for 2 boreholes from 8, where the subsidence rate is about 0.08 km/Ma. The difference in another 3 boreholes is less than 15 K/km. A poor fit is found for 3 boreholes in German basins which subsided or began to subside during the Upper Carboniferous. The difference in the methods probably results from the simplification and from differences in the principles used in the two methods.

Introduction

The paleogeothermal interpretation of the degree of coalification was first worked out by HUCK & KARWEIL (1955). In the seventies, when the reflectance method was introduced in nearly all coal petrological laboratories, a lot of data were available for interpretation. Some methods were developed describing an empirical relationship between vitrinite reflectance and temperature, as well as duration of its exposure (e.g., LOPATIN, 1971; BOSTICK, 1973; HOOD et al., 1975; BUNTEBARTH, 1978).

All methods were developed from data of distinct basins in which local effects are involved which cannot be considered with simple mathematical methods, e.g. different plant remains and different environmental properties.

This paper aims to estimate the paleogeothermal gradient of vitrinite reflectance data from Australia and from the F.R. of Germany applying the methods reported by BUNTEBARTH (1982). The geological age of the sedimentary sequences ranges from Upper Carboniferous to Tertiary.

Applied methods

The method given by BUNTEBARTH (1978) yields the paleogeothermal gradient from

(1) $\quad Rm^2 = 0.00116 \exp(0.068 \, dT/dz) \int_0^t z(t^*) \, dt^*$

where Rm = vitrinite reflectance, t = time, $z(t^*)$ = burial depth of the sediment under consideration in t^*, dT/dz = temperature gradient during the subsidence (supposed to be constant).

If Eq. (1) is applied to basins with a constant temperature gradient and a constant subsidence rate which might approximate many geological cases, the integrated subsidence history becomes:

(2) $\quad \int_0^t z(t^*) \, dt^* = z^2/2V$

where z = maximum depth of burial of the layer at the time t, and V = subsidence rate.

From Eq. (1) and Eq. (2) follows:

(3) $\quad Rm = A\,z$

where A = constant is equal to:

(3) $\quad A = 0.024/\sqrt{V} \exp(0.034 \, dT/dz)$

For the simplified assumption of constant subsidence and burial rate from the sedimentation of a particle to the maximum depth of burial, a linear increase of vitrinite reflectance Rm with depth z follows. Eq. (3) implies that the geothermal gradient can be estimated directly from the linear relation between vitrinite reflectance and depth. The slope (A) of the Rm versus z curve yields the temperature gradient:

(4) $\quad \frac{dT}{dz}(K/km) = 29.4 \ln(41.5 \, A\,V)$

using the dimensions A (%/km) and V (km/Ma).

Eq. (3) implies that $Rm = 0$ at $z = 0$. Furthermore, the coalification must be preorogenic, i.e. neither standstill, uplift, or erosion occurred during the sedimentation, nor has the organic matter undergone a higher reaction temperature in the time subsequent to the deepest burial. In order to take into account the effect of moving water in sedimentary basins and also some other effects, the influence of which can hardly be estimated, i.e. a standstill or uplift with subsequent erosion, a constant c has been introduced (BUNTEBARTH & TEICHMÜLLER, 1982) which expands Eq. (2) to

$$\text{(5a)} \quad \int_0^t z(t^*)dt^* = \int_{t_1}^t (\tilde{z}(t^*)+z_e)dt^* + c$$

$$\text{(5b)} \quad \qquad\qquad = (\tilde{z}+z_e)^2/2V + c$$

where z is the observed depth, and z_e is the thickness which has been eroded. Eq. (5) still approximates the subsidence history with a constant rate.

The method of MIDDLETON (1982) was originally proposed to explain the logarithmic increase in vitrinite reflectance with depth described by DOW (1977). The method expresses the empirical nomogram of SHIBAOKA & BENNETT (1977), which is determined from a database of over 1000 vitrinite reflectance measurements on Australian coals, as an integral equation relating vitrinite reflectance to temperature and time. The equation is

$$\text{(6)} \quad Rm^a = b^* \int_0^{t_1} \exp[cT(t)]dt$$

where the initial reflectance of vitrinite before burial (in the peat state, approximately 0.2 %), is involved in b*, T(t) is the temperature history of the coal, and the empirical constants a = 5.5, b* = $2.8 \cdot 10^{-6}$ and c = 0.068.

If the geothermal gradient is constant, T(t) can be replaced by

$$T(t) = \frac{dT}{dz} z(t)$$

and in the case of a coal seam subsiding at a constant rate

$$V = \frac{dz}{dt},$$

the relation in Eq. (6) reduces after integration to the simple expression:

$$\text{(7)} \quad Rm = B \exp[0.01236 \, dT/dz \, z]$$

where B is a constant, z is the depth of burial and dT/dz is the geothermal gradient (assumed constant). Eq. (7) is equivalent to the logarithmic expression of DOW (1977).

Under the assumptions of a constant subsidence rate and a constant geothermal gradient, the geothermal gradient can be estimated by (i) matching a curve of the form $R_m = B \exp(bz)$ to observed vitrinite reflectance versus depth data by least squares regression or similar method. Then, b from the matched curve is identified with the exponent of Eq. (7):

$$b = 0.01236 \, dT/dz,$$

so that the temperature gradient can be calculated

(8) $\quad \dfrac{dT}{dz} = 80.9 \, b$

MIDDLETON (1982) demonstrated that, where erosion had taken place after burial, Eq. (7) was still valid but the value of the constant B was multiplied by the factor $\exp[0.01236 \, dT/dz \, z_e]$, where z_e is the thickness of layers which has been eroded.

Results

The vitrinite reflectance data of 8 boreholes are used to estimate temperature gradients using the two methods. Four boreholes are from the F.R. of Germany and four boreholes are from Australia. The range of reflectance varies between 0.16 and 1.84 % Rm. Table 1 summarises these data with source references. The BUNTEBARTH method is applied using Eq. (3) to match the data and Eq. (4) to calculate dT/dz. The MIDDLETON method is applied using $Rm = B e^{bz}$ to match the data and Eq. (8) to calculate dT/dz.

Figs. 1 to 8 show the application of both methods to the 8 boreholes. Two curves are matched by least squares regression to each data set. One curve (——) in each case is the application of BUNTEBARTH's method and the other (---) is the application of MIDDLETON's method. As expected from Eq. (3) and Eq. (7), the curves are linear and exponential, respectively.

The results of the comparison are shown in Table 2. The tabulated information presents (i) average subsidence, used for V in Eq. (3), (ii) calculated dT/dz for both methods and (iii) either the recent (measured) or expected temperature gradient for each borehole.

A good fit between the two methods is found for Sandhausen 1 (6 K/km difference) and Jupiter 1 (2 K/km difference). A worse fit is found for the other 3 boreholes: Pecten 1A (13 K/km difference), Bass Basin (14 K/km difference) and Boomi 1 (12 K/km difference). A poor fit is found for 3 boreholes in basins which subsided or which began to subside during the Upper Carboniferous: Urach 3 (31 K/km difference), Eiweiler-Vogelsborn (26 K/km difference), and Nordlicht Ost 1 (29 K/km difference).

Table 1. Vitrinite reflectances and geological data of the boreholes used for comparison

Borehole	Vitrinite reflectance Rm (%)	Number of Rm values evaluated	Stage (mainly)	Total burial depth of the samples (km)	References
Sandhausen 1/FRG	0.16 - 0.73	35	Tertiary	0.64 - 2.9	TEICHMÜLLER & TEICHMÜLLER, 1981
Bass Basin/Australia	0.20 - 1.1	11	Tertiary	0.45 - 3.2	KANTSLER et al., 1978
Pecten 1A/Australia	0.45 - 0.64	6	Cretaceous, Tertiary	1.74 - 2.83	MIDDLETON & FALVEY, 1983
Urach 3/FRG	0.5 - 1.84	22	Triassic, Upper Carboniferous	0.6 - 2.1*	BUNTEBARTH et al., 1979
Jupiter 1/Australia	0.27 - 1.36	15	Jurassic, Triassic	1.04 - 3.98**	BARBER, 1982
Boomi 1/Australia	0.36 - 0.51	6	Jurassic, Triassic	0.62 - 1.59	MIDDLETON & RUSSELL, 1981
Eiweiler-Vogelsborn/FRG	0.74 - 0.95	6	Upper Carboniferous	1.14 - 1.69*	TEICHMÜLLER et al., 1983
Nordlicht Ost 1/FRG	0.70 - 1.54	37	Upper Carboniferous	0.77 - 1.97*	BUNTEBARTH et al., 1982

*) some eroded thickness has been considered
**) depth below sea bottom

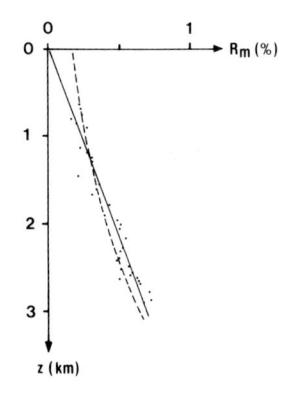

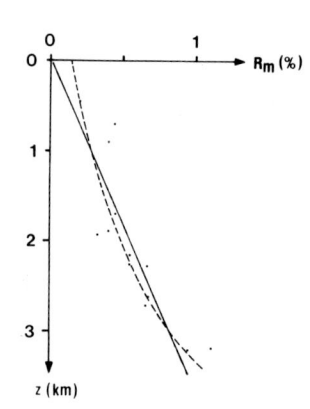

Fig. 1. Vitrinite reflectance versus depth for the borehole Sandhausen 1/FRG after TEICHMÜLLER (1970). The linear curve (——) results from the method by BUNTEBARTH and the exponential curve (---) from that by MIDDLETON.

Fig. 2. Vitrinite reflectance versus depth for the Bass Basin/Australia after KANTSLER et al. (1978). The linear curve (——) results from the method by BUNTEBARTH and the exponential curve (---) from that by MIDDLETON.

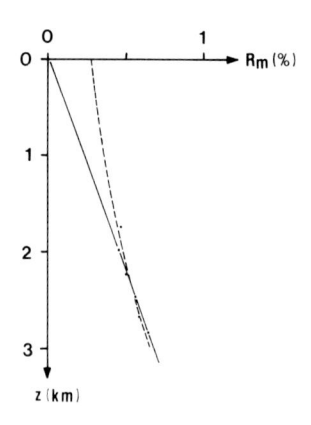

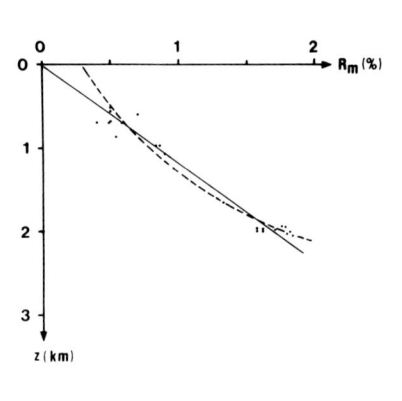

Fig. 3. Vitrinite reflectance versus depth for the borehole Pecten 1A/Australia after MIDDLETON & FALVEY (1983). The linear curve (——) results from the method by BUNTEBARTH and the exponential curve (---) from that by MIDDLETON.

Fig. 4. Vitrinite reflectance versus depth for the borehole Urach 3/FRB after BUNTEBARTH et al. (1979). The linear curve (——) results from the method by BUNTEBARTH and the exponential curve (---) from that by MIDDLETON.

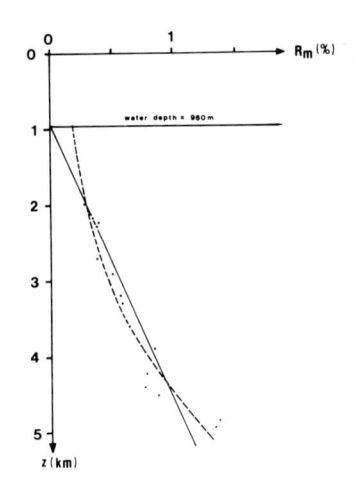

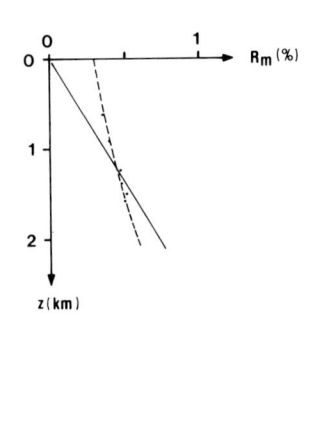

Fig. 5. Vitrinite reflectance versus depth for the borehole Jupiter 1/Australia after BARBER (1982). The linear curve (———) results from the method by BUNTEBARTH and the exponential curve (---) from that by MIDDLETON.

Fig. 6. Vitrinite reflectance versus depth for the borehole Boomi 1/Australia after MIDDLETON & RUSSELL (1981). The linear curve (———) results from the method by BUNTEBARTH and the exponential curve (---) from that by MIDDLETON.

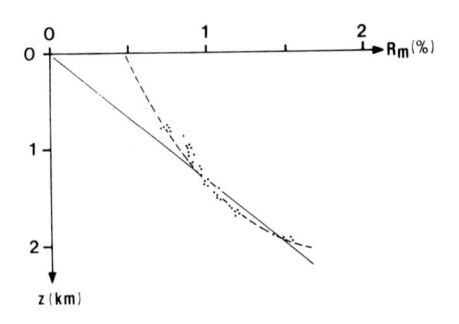

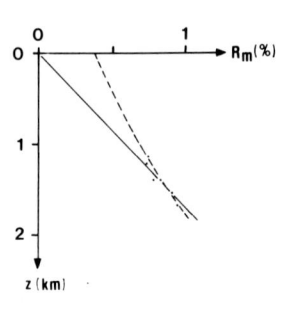

Fig. 7. Vitrinite reflectance versus depth for the borehole Eiweiler-Vogelsborn/FRG after TEICHMÜLLER et al. (1983). The linear curve (———) results from the method by BUNTEBARTH and the exponential curve (---) from that by MIDDLETON.

Fig. 8. Vitrinite reflectance versus depth for the borehole Nordlicht Ost 1/FRG after BUNTEBARTH et al. (1982). The linear curve (———) results from the method by BUNTEBARTH and the exponential curve (---) from that by MIDDLETON.

Table 2. Estimated paleotemperature gradients for 8 boreholes using BUNTEBARTH's and MIDDLETON's method

Basin/Geological Age (Ma)	Borehole	Average Subsidence Rate (km/Ma)	Temperature gradient (K/km) after			Recent	References
			BUNTEBARTH (B)	MIDDLETON (M)	Difference B-M		
Upper Rhine Graben/ FRG, 0-37	Sandhausen 1	0.08	29	35	-6	42*	TEICHMÜLLER, 1970; BUNTEBARTH, 1978; TEICHMÜLLER & TEICHMÜLLER, 1979
Bass Basin/Australia 0-55	Composite	0.076	35	49	-14	35	KANTSLER et al., 1978
Southern Continental Margin/Australia 0-110	Pecten 1A	0.026	12	25	-13	20-30	MIDDLETON & FALVEY, 1983
Swabian Alb/FRG 140-290	Urach 3	0.015	43	74	-31	39	BUNTEBARTH et al., 1979; BUNTEBARTH & TEICHMÜLLER, 1982; HAENEL & ZOTH, 1982
Exmouth Plateau/off-shore W Australia 180-220	Jupiter 1	0.08	35	37	-2	30-40**	BARBER, 1982
Gunnedah Basin/ Australia 210-100	Boomi 1	0.015	18	30	-12	20-30**	MIDDLETON & RUSSELL, 1981
Saar-Nahe Basin/ FRG, 280-292	Eiweiler-Vogelsborn	0.227	71	45	+26	29	HEDEMANN, 1976; BUNTEBARTH, 1983; TEICHMÜLLER et al., 1983
Ruhr Basin/FRG 297-308	Nordlicht Ost 1	0.19	76	47	+29	43	TEICHMÜLLER, 1973; BUNTEBARTH et al., 1982; WEBER, 1984

*) Mean value from bottom hole temperature **) expected values

With regard to the BUNTEBARTH method, if the more detailed subsidence history, Eq. (5), is applied instead of the simplification with V = constant, some deviation appears in the results. The high Upper Carboniferous values are about 10 % less than given in Table 2 (BUNTEBARTH et al., 1982), and the temperature gradient of the Sandhausen borehole is about 10 % higher (BUNTEBARTH, 1978).

The vitrinite reflectance versus depth does not give a proper fit to the data of the borehole Boomi 1, if Eq. (3) is applied (Fig. 6). The data would give a better approximation, if some overburden has been eroded.

In the case of Boomi 1, MIDDLETON's method also indicates that some erosion has occurred, which results from a surface reflectance greater than 0.2 % Rm (see Fig. 6). The amount of erosion is intrinsic to the factor B in Eq. (7).

Conclusions

The two methods of estimating paleogeothermal gradient described herein are concordant only under specific conditions. The discrepancy between the two methods appears to arise from two sources: (i) the methods were devised empirically to describe sedimentary basins with different tectonic histories, and (ii) the method of MIDDLETON allows coalification to increase at a slow rate with constant reaction temperature.

The method of MIDDLETON (1982) was based upon an empirical correlation between vitrinite reflectance, burial temperature and burial time (SHIBAOKA & BENNETT, 1977) for Australian sedimentary basins. This correlation is strongly influenced by slowly subsiding Permian (Gondwana) coal basins which have experienced a time of burial of the order of 50 - 100 Ma before uplift moved them closer to the surface (or exposed them).

BUNTEBARTH (1978, 1979) based his original correlation on the young basins of southern Germany. His method demonstrably works well also for rapidly subsiding Carboniferous basins. Therefore, one possible reason for the discordance of the two methods under certain conditions is that they were devised from basins that had experienced different tectonic histories.

The results given in Table 2 demonstrate that the temperature gradients are in good agreement for data from recently subsiding basins, where subsidence ceased at present, and also from basins with a moderate subsidence rate. The estimates are in close accord with expected values of the temperature gradient. The method by MIDDLETON appears to give higher values than the method by BUNTEBARTH with exception of the Upper Carboniferous German basins. Both methods account for long periods of burial with or without very low subsidence in different ways. The difference is due to the fact that MIDDLETON allows coalification to increase at a slow rate with constant reaction temperature:

(9a) $\quad Rm^{5.5} = \text{const.} \int_0^t \exp(0.068\ T)\ dt^*$

(9b) $\quad\quad\quad = \text{const.} \exp(0.068\ T) \int_0^t dt^*$

This is suggested by the nomograms of BOSTICK et al. (1979) and HOOD et al. (1975). However, BUNTEBARTH considers only a kinematic process, where the subsidence rate is not equal to zero.

Time dependence of vitrinite reflectance is a major unknown in thermal modelling applications. This uncertainty is demonstrated by (i) the experimental work of BOSTICK (1973), which shows that coal needs to be heated at 400° C for a month to achieve a vitrinite reflectance of 2 %, (ii) the nomogram of HOOD et al. (1975), devised from extensive sedimentary basin analysis, which suggests heating coal for 10 Ma at 200°C to achieve 2 % vitrinite reflectance, and MIDDLETON (1982) who suggests heating coal for 40 Ma at 200° C to achieve 2 % vitrinite reflectance. In each case, specific samples justify the conclusions. In each case, the specific samples have a different chemistry and thermal-burial history. Clearly, extrapolation of the behavior of vitrinite reflectance to temperature throughout geological time is strongly related to the chemistry of the coal and the thermal-burial history of the coal basin. Both empirical methods herein described work well within the geological regions for which they were developed, but only under certain strict conditions outside those regions.

METHODS FOR PALEOTEMPERATURE ESTIMATION USING VITRINITE REFLECTANCE DATA : A CRITICAL EVALUATION

VETŐ*, I. and P. DÖVÉNYI**

* Hungarian Geological Survey
Népstadion út 14, H-1143 Budapest

** Geophysical Department, Eötvös University
Kun Béla tér 2, H-1083 Budapest

Abstract

Several empirical models were published to describe the relation between burial history, geothermal conditions and the maturity of organic matter in sedimentary basins. Four widely accepted models (LOPATIN, BOSTICK, HOOD, WAPLES) are compared on the basis of 60 measured vitrinite reflectance profiles from different parts of the world. All the reflectance data were taken from the literature together with burial histories and rock temperature.

Since a pure statistical approach of the errors would have little sense, we tried to investigate the problem by the study of the following relations:

- errors vs. type of burial history
- errors vs. level of measured reflectance.

In this way we are able to discuss the application limits of the studied methods. Our main conclusion is that satisfactory results can be obtained by the HOOD and WAPLES method in a limited reflectance range ($R_o = 0.4 - 1.0$ %).

1. Introduction

A number of models were proposed to describe quantitatively the relationship between thermal history and organic maturity since KARWEIL published the first attempt in 1956. All the models have the following assumptions:

- The maturation of organic matter is controlled by temperature alone.
- The maturation process is irreversible.
- The same thermal history yields the same maturity.
- The higher the temperature the higher the maturity if all the other conditions are the same.

However, the models strongly differ in the way in which time is involved in the maturation; some models completely neglect its significance.

The interest of petroleum geologists and geothermal specialists was aroused by the possibility of maturity or paleotemperature estimation. Some of the methods have been applied widely in different basins. The question now arises as to the reasons why a given model was developed. In the literature no answer can usually be found, so we conclude the choices were made either by sympathy, or by the lack of knowledge of other methods. Relatively little effort was made to compare the advantages and the limitations of the different models (KETTEL, 1981; VETÖ, 1981; VETÖ et al., 1984). The aim of our paper is to provide this comparison.

2. Comparison of the models

In the following we use the vitrinite reflectance to describe the level of maturity in the same way as the methods considered. With increasing maturity the vitrinite becomes more and more anisotropic and both its maximum reflectance and its average reflectance (Rm) become higher. These two values correspond strictly with each other, however, at high reflectance values (i.e. above 4 % Rm) the maximum reflectance should be applied (TEICHMÜLLER, 1982).

2.1 The common basic assumptions

The activation energy required for coalification can be assured not only by heating but also by natural radioactivity (TEICHMÜLLER & TEICHMÜLLER, 1958). Since the increase of R_O due to the effect of radioactivity is of importance only if the embedding rock has a remarkable concentration of uranium and/or thorium. The role of radioactivity in coalification is statistically insignificant. On the basis of laboratory experiments, TROFIMUK et al. (1983) suppose that the seismic energy can directly activate the coalification. Since this phenomenon is not yet proved on geological examples we shall neglect it. So we accept temperature as the practically only activator of coalification.

Obviously the irreversibility of the R_o increase cannot be proved by measurements but we have no serious reason to doubt it.

The third common assumption, namely a unique response of vitrinite to the same thermal history, and the fourth one that the higher the temperature, the higher the maturity, seem to be in contradiction with a lot of observations. Here we report some of them:

HUTTON et al. (1980) studying torbanitic oil shales have found a decrease in $R_o max$ from 0.6 - 0.75 % to 0.3 % with a parallel increase of the alginite fraction from 0 % to 80 % in very neighbouring samples. Coals and pelitic rocks are often characterized by higher $R_o max$ than the neighbouring sandy ones. The difference can go up to 30 %

(BOSTICK & FOSTER, 1975; KÜNSTNER, 1974). These observations were done in coal measures with R_0max varying between 0.5 % and 1.9 %. BLANQUART & MERIAUX (1975) have found in the Nord coal basin (NE France) that the R_0max of coaly inclusions increases from 1 % to 1.6 % with a corresponding increase of their thickness from 20 µm to 300 µm. CARRETA & WOLF (1980) studied thin Gondwana coal seams from Brasilia. With a dense sampling they recognized a decrease in R_0 of up to 0.4 % going upward in a seam of 2.2 m in thickness (the average R_0 of the seams was 1.1 %). BONES et al. (1972) described similar phenomena in Carboniferous coals of NE England. NIKOLOV (1974) reported a decrease in R_0 of up to 0.15 % 1 - 2 m below the intraformational erosion surfaces in the Carboniferous coal measures of Dobrudja (NE Bulgaria). BONES et al. (1972) report a 0.1 % higher R_0 in a seam which is covered by claystone than in the same seam covered by sandstone (the average R_0 of the seam varies between 0.6 and 1.0 %). NEWMAN & NEWMAN (1982) studied a coal deposit of the Southern Island (New Zealand). The five seams developed in a 200 m thick tectonically undisturbed sequence show the following R_0 values upwardly: 0.6 %, 0.67 %, 0.8 %, 0.84 %, 0.92 %, 0.74 %, 0.73 %.

The evaluation of the frequency of these phenomena observed mostly in coal is beyond the scope of this paper. But it shows that the response of vitrinite to a certain thermal history can vary widely. This fact has to be taken into account in the discussion of our topic.

2.2 The models for comparison

We do not intend to evaluate every published model (KARWEIL, 1956; LOPATIN, 1971; BOSTICK, 1973; CORNELIUS, 1975; HOOD et al., 1975; KARPOV et al., 1975; TISSOT & ESPITALIÉ, 1975; LOPATIN, 1976; AMMOSOV et al., 1977; SHIBAOKA & BENNETT, 1977; WAPLES, 1980; BUNTEBARTH, 1982). Our ambitions are limited to the four most widespread ones (BOSTICK, HOOD et al., LOPATIN, WAPLES). The AMMOSOV et al. model neglects the role of time in coalification and although very popular in soviet literature, is not taken into consideration because it was refuted convincingly by VELEV et al. (1979).

HOOD et al. (1975) proposed to calculate the maturity from the maximum temperature (T_{max}) and from the "effective heating time" (t_{eff}). This latter is the time which the organic matter spent in the temperature interval of (T_{max}, $T_{max} - 15°$ C).

LOPATIN (1976) divided the thermal history above 50° C into temperature intervals of increasing length and multiplies the time which an organic particle spent in these intervals by 2^n (n = 0,1,2...) to get elementary heat impulses. The vitrinite reflectance is assumed to be a given function of the summarized elementary heat impulse ($\Sigma\tau$).

WAPLES (1980) accepted LOPATIN's (1972) original idea, divided the thermal history for equal time intervals and gave an improved empirical relationship for the summarized heat impulses (Time-Temperature Index, TTI) and vitrinite reflectance.

BOSTICK (1973) derived an "R_o" value from the maximum temperature and from the time the organic particle spent at temperatures higher than 20° C. The actual vitrinite reflectance was calculated by multiplication of "R_o" by the ratio of shaded area to the area of the time-temperature rectangle on Fig. 1.

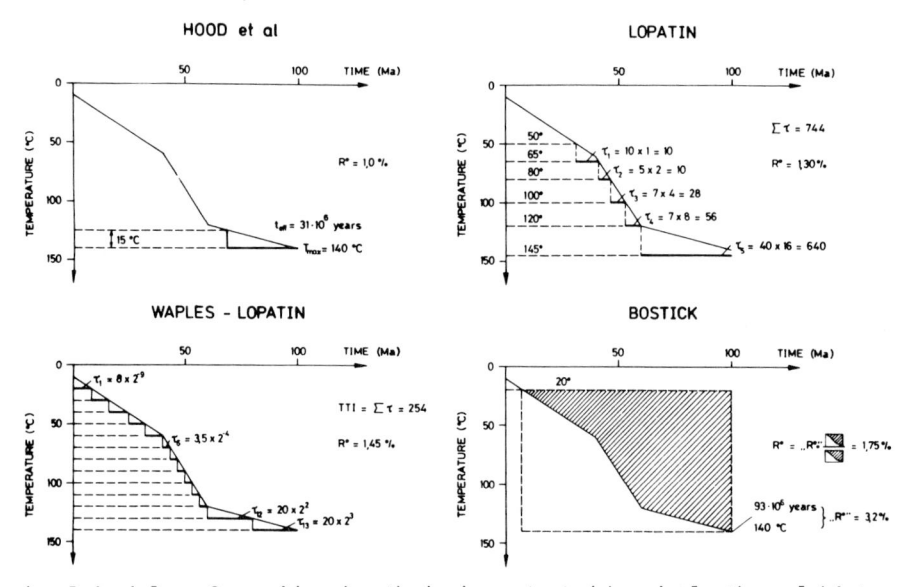

Fig. 1. Principles of considered methods demonstrated by simple thermal history

There are some contradictions in the methods. Thermal histories can be created for both the BOSTICK and the HOOD methods showing R_o drop as a result of temperature increase, but most of these artificial thermal histories have no practical importance. There are counter-arguments from a geochemical point of view against the LOPATIN and the WAPLES models, too. All in all, these empirical methods can be qualified only by comparison of modeled and measured vitrinite reflectances.

2.3 Data base of comparison

We have collected temperature values, burial histories and more than 500 vitrinite reflectance data of 60 boreholes from the literature. Their locations are shown on Fig. 2. The names of sites and basins are listed in the appendix. Most of the literature report tabulated R_o values and some of them present only best fit R_o-depth curves. Some obviously erroneous R_o values were not used. If the surface (or sea bottom) temperature was not given we have estimated it from the Pergamon World Atlas (1968) or from the Atlas of Oceans (1977).

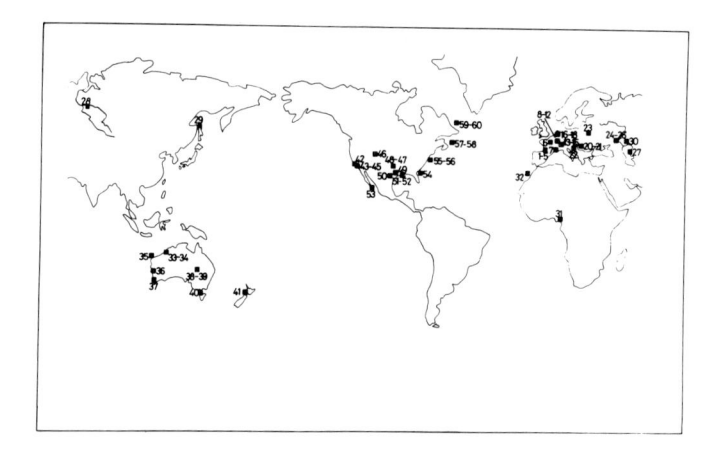

Fig. 2. Location of boreholes from which the data were taken

If the burial history was presented explicitly, we used that, if not, we constructed it from the given stratigraphic data. In this case the sedimentation rate was taken to be constant between two consecutive stratigraphic data. If there was a stratigraphic hiatus but no further information, we did the following: The first half of the corresponding time interval was taken into account with the previous sedimentation rate and the second one with an equal erosion rate. Correction for compaction was not applied. If the papers did not report absolute geological ages we used the dates in VAN EYSINGA's (1972) compilation.

The correct comparison of models may be hindered by the errors of input data:

- reflectance measurement
- temperature measurements
- burial history
- differences between the present and the former geothermal gradient
- differences between the present and the former surface temperature.

We are unable to avoid the inaccuracy of R_o and of temperature measurements. The errors in burial history are the smallest in the case of continuously subsiding basins.

Constant geothermal gradient in space and time is supposed for all the models we are discussing. This assumption is in contradiction with many facts of which the most important are listed below:

- Different rocks have different heat conductivities.
- Different sediments compact at different rates.
- Fast sedimentation has a cooling effect.
- Convective heat transport (volcanic activity, water circulation) may disturb the thermal equilibrium.
- There are long-term changes in surface (ocean floor) temperature.
- Mantle heat flow is not constant in time.

In spite of these facts it may be supposed that in a continuously subsiding basin each organic particle reaches the highest temperature at present. It is obvious that different heat conductivities, sediment compaction rates and the cooling effect of fast sedimentation (STEGENA & DÖVÉNYI, 1983) may hinder but not stop or reverse the temperature increase.

This is not the case near volcanic areas, but this effect is usually negligible on a regional scale because the zone of influence is very limited (HORVÁTH et al., in this volume). Localized and slow circulation of water can also be neglected in sedimentary basins filled predominantly with marls and sandstones.

On tectonically quiet areas (e.g. platforms) a quasi-constant mantle heat flow can be supposed for a long time. For a detailed study it can be stated that the actual rate of change in mantle heat flow in these areas is controlled by the total heat loss of the Earth. This loss decreases with time. TURCOTTE (1980) suggested that it was 10 to 12 % higher 500 Ma ago. If this is taken into account in a continuously subsiding basin with a present depth of 5 km we can conclude that the thermal history of each particle shows continuous increase unless the subsidence rate is less than 1 m/Ma. The average low subsidence velocity of the sedimentary basins is much higher (~20 m/Ma according to GRETENER & CURTIS, 1982).

There are quantitative methods to describe subsidence and thermal history of intracontinental basins and shelves which were generated by extensional tectonics (McKENZIE, 1978). The calculation (McKENZIE, 1981) shows monotonous temperature increase of sediment particles deposited after the beginning of extension.

The variation of surface temperature does not effect practically the thermal conditions of deeper (>1 km) strata.

We may conclude that continuously subsiding basins give the best opportunity for the comparison of the different methods.

3. The comparison of the models

We estimated the errors of the different models by calculating relative differences between the measured and predicted vitrinite reflectance values. The computation of predicted values was carried out by keeping precisely the prescription of the four authors with the exception of the Hood method, which was used with a lower R_0 limit of 0.4 % instead of the original one of 0.5 %. The results are represented in relative deviance (%) vs. relative frequency (%) diagrams for each method. To prepare these diagrams we divided the 0.2 % - 6.0 % vitrinite reflectance range into 0.2 % long intervals. We included into the calculation only the average deviance values from each interval of one borehole. This way we could eliminate the undesirable influence of sites with an extremely large number of vitrinite reflectance determinations.

3.1 Continuously subsiding basins with vitrinite reflectance below 2 %

The average errors obtained are presented in histograms in Fig. 3. We note the measured R_0 values cover only the range of 0.2 % - 2.0 %. The HOOD (H) and WAPLES (W) models give satisfactory evaluation in two-thirds of the cases.

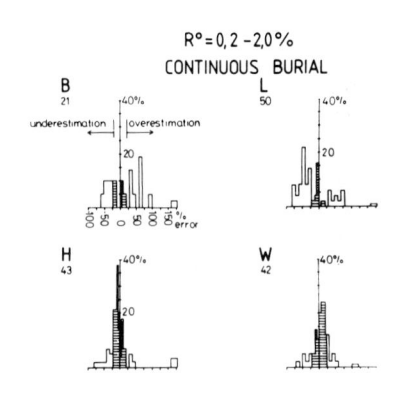

Fig. 3. Histograms of differences between the calculated and measured vitrinite reflectances in continuously subsiding basins. The hatchured columns show the area of correct estimations. It also shows the number of data considered.

We consider an evaluation is satisfactory if the error is less than ± 20 %. This limitation is conservative enough since a 30 % relative difference between R_0 values of rocks with identical thermal history is not too rare. On the contrary, the BOSTICK (B) and LOPATIN (L) models work correctly only in one-fifth of the cases.

Since the measured R_0 values do not uniformly cover the 0.2 - 2.0 % range and the type of chemical reaction changes as vitrinite passes through this R_0 range (van KREVELEN, 1963), we have also plotted similar histograms for narrower R_0 ranges (Fig. 4). In the beginning of coalification (R_0 = 0.2 - 0.4 %) the W model overestimates and the L model underestimates (the two other methods do not work in this R^0 range). In the 0.4 - 0.8 % range the H and W models work well (the evaluations are

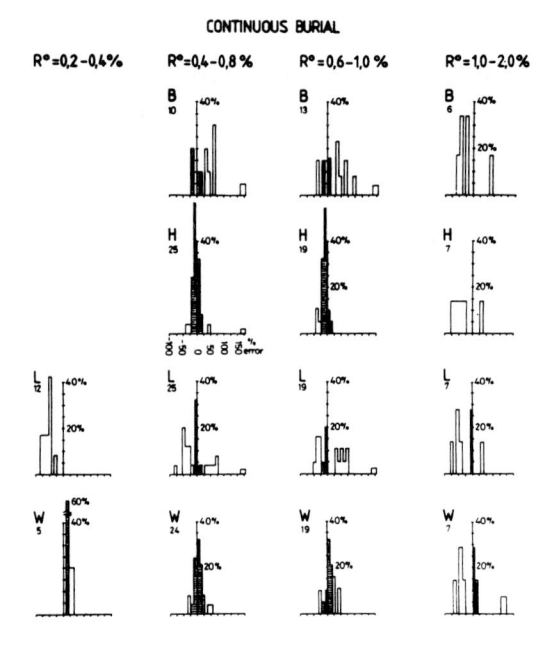

Fig. 4. Histograms of differences between the calculated and measured vitrinite reflectances for continuously subsiding basins in different reflectance intervals. The hatchured columns show the area of correct estimates. It also shows the number of data considered.

satisfactory in 84 % and 76 % of the cases). The B and L models are characterized by over- and underestimations respectively. In the partly overlapping 0.6 - 1.0 % range we have obtained a similar picture, only the overestimating tendency of the W model is enhanced. It is to be noted that with the H model the weak mild underestimation is enhanced by the increase of measured R_o.

We conclude that the H and W models describe correctly the dependence of R_o from thermal history in the R_o range of 0.4 - 1.0 % which correspond more or less to the oil window. The B and L models do not work well in this case. The small number of evaluations in the R_o range over 1.0 % does not permit judgement of how the models work there.

3.2 Highly matured basins

For the maturity range of R_o 1.0 % the comparison can be made only by also using the more uncertain thermal histories of discontinuously subsiding basins. Fig. 5 shows histograms of the 4 models, separately for R_o range of 1.0 - 2.0 % and of 2.2 - 6.0 %. In the range of 1.0 - 2.2 % not one of the models works well, even the B model only gives satisfactory estimations in every second case. In the range of 2.2 - 6.0 % the B model seems to work well, unlike the others. The H model gives only strong underestimations.

Taking into account the type of burial history the picture changes. In Fig. 6 the errors are presented in groups characterized by a similar type of burial. It is very

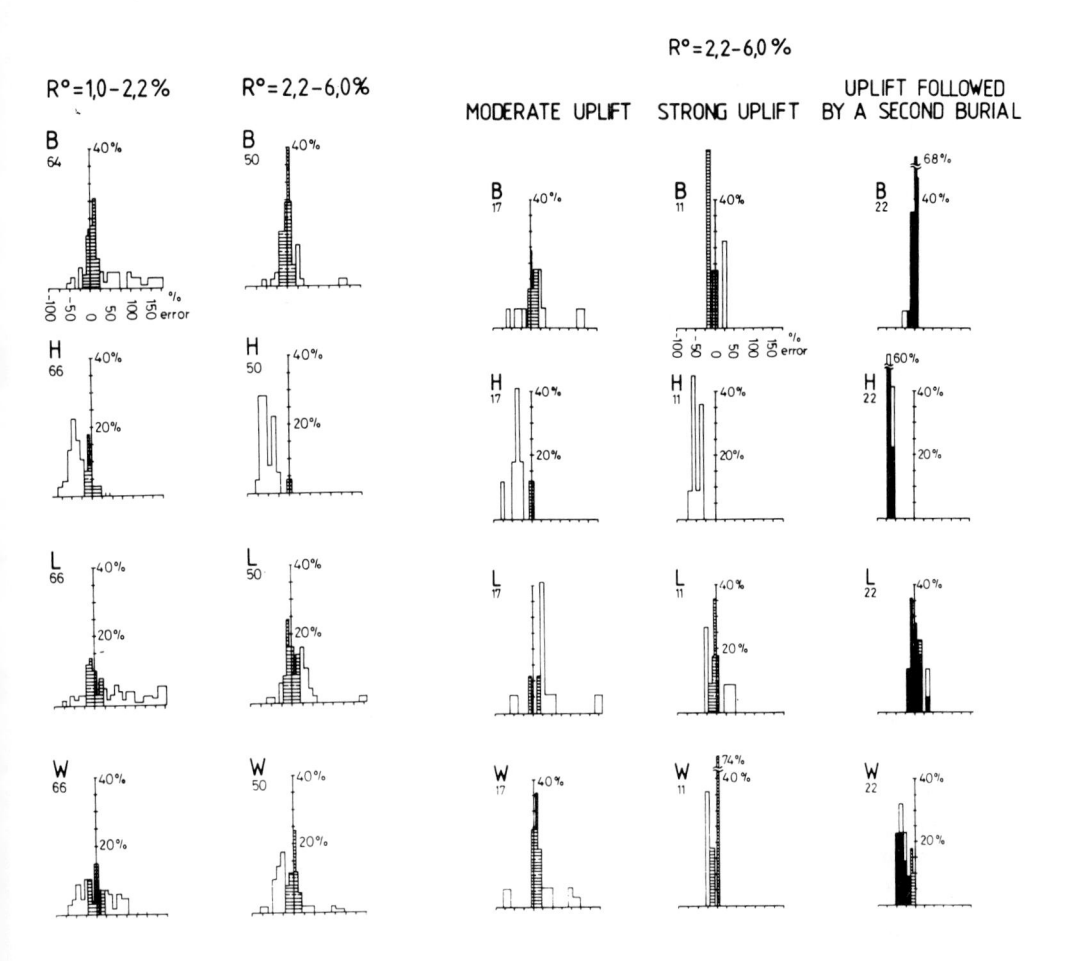

Fig. 5. Histograms of differences between the calculated and measured vitrinite reflectances for all types of basins in the reflectance intervals of 1.0-2.2 % and 2.2-6.0 %. The hatchured columns show the area of correct estimations. The number of data considered is also shown.

Fig. 6. Histograms of differences between the calculated and measured vitrinite reflectances for different type of basins in the reflectance range of 2.2-6.0 %. The black columns are derived from the data of Münsterland 1 borehole. The hatchured columns show the area of correct estimations. The number of data considered is also shown.

remarkable that the best evaluations of the B model were obtained in basins with two periods of burial, separated by a long period of uplift, i.e. with very uncertain thermal history.

On the histograms we marked the part representing the Münsterland 1 borehole. The evaluated R_o values for this borehole form a significant part of the satisfactory ones. The Münsterland 1 is located north of the Ruhr basin (FRG) where the rank of the coal of the same age (Upper Carboniferous) was reached before the Permian in a thermal regime characterized by very high (70 - 80° C/km) geothermal gradients (TEICHMÜLLER & TEICHMÜLLER, 1979).

The coal deposits of the Münsterland 1 experienced two intensive burials, i.e. during Upper Carboniferous and Upper Cretaceous times. The very good R_o estimations of the B and L models were calculated with the recent geothermal gradient (40° C/km) and with more than 100 Ma as main acting time. We suppose the two errors of the postulated thermal history more or less counterbalance each other.

It is difficult to estimate the correctness of models in the higher R_o ranges because we cannot separate the errors due to the model and due to the postulated thermal history.

4. How to use the models for paleogeothermal reconstruction

The scarcity of data and/or the uncertainty of the thermal histories does not permit judgement of the usefulness of the 4 models at R_o values higher than 1 % or lower than 0.4 %. In the range of 0.4 - 1.0 % the H and W models describe correctly the relation between the thermal history and vitrinite reflectance.

The mild overestimation of W method is probably the error of the W model itself, since the decrease of geothermal gradient expected generally during the burial would result in underestimations. In the underestimating tendency of H model we are unable to differentiate the errors due to the model and due to decrease of geothermal gradient.

We have to conclude that the use of H and W methods is limited in paleogeothermal reconstruction. In the 0.4 - 1.0 % R_o range, correct estimation proves that the present geothermal gradient is the same as it was in the period of maximum heating. In continuously subsiding basins this signifies the recent temperatures as maximum ones. In the case of strong underestimation (more than 20 %) we have to suppose a very fast decrease in heat flow during the burial.

YORATH & HINDMANN (1983) present an example for underestimation. In 2830 m depth of the Murrelet borehole, Queen Charlotte basin, British Columbian offshore, a vitrinite reflectance of 0.95 % and a temperature of 85° C were measured. The authors assume a much higher paleo-heat flow caused by a rifting older than 17 Ma.

Strong overestimation probably indicates a very young heating event, but we have no illustrative example for strong overestimation in continuously subsiding basins.

Appendix

Well	Basin or location	References
1. Lacq. 104	Aquitaine	BRGM, 1974, Géologie du Bassin d'Aquitaine
2. Nassiet 1		Robert, P., 1976, Bull. CRP 10.1 271-285
3. Pont d'As 1		
4. Rousse 1		Tissot, B., Espitalié, J., 1975, Revue IFP XXX. 5. 743-777
5. Lons 1	Aquitaine	BRGM, 1974, Géologie du Bassin d'Aquitaine Correia, M., Peniguel, G.,1975, Bull. CRP 9.2 99-127 Le Tran, K., 1971, Bull. CRP 5.2 321-332
6. Essises 1	Paris	Tissot, B. et al., 1974, AAPG Bull. 58.3 499-506 BRGM, 1969, Carte Géol. à 1/50.000 CHATEAU THIERRY
7. Pierrefeu 1	Rhône	Dunoyer de Segonzac, G., 1969, Mém. Carte Géol. Als. Lorr. 29
8. Harthausen 1	Rhine Graben	Teichmüller, M., Teichmüller,R., 1979, Fortschr. Geol. Rheinl. u. Westf. 27: 109-120
9. Hähnlein West 2		
10. Landau 2		Teichmüller, M., 1979, Fortschr. Geol. Rheinl. u. Westf. 27: 19-49
11. Sandhausen 1	Rhine Graben	Teichmüller, M., Teichmüller,R., 1979, Fortschr. Geol. Rheinl. u. Westf. 27: 109-120
12. Urach 3	Swabian Alb	Buntebarth, G. et al., 1979, Fortschr. Geol. Rheinl. u. Westf. 27: 183-199
13. Anzing 3	Molasse	Teichmüller, M., Teichmüller, R., 1975, Geol. Bavar. 73: 123-142
14. Miesbach 1	Molasse	Jacob, H., Kuckelhorn, K., 1977, Erdöl-Erdgas Z. 93.4 115-124
15. Staffelsee 1	Molasse	Jacob, H. et al., 1982, Erdöl und Kohle, 35.1. 511-518

Well	Basin or location	References
16. Münsterland 1	Westphalia	Teichmüller, M., Teichmüller, R., 1979, Fortschr. Geol. Rheinl. u. Westf. 27: 109-120 Richwien, J. et al., 1963, Fortschr. Geol. Rheinl. u. Westf. 11: 9-18 Hedemann, H.A., 1963, Fortschr. Geol. Rheinl. u. Westf. 11: 403-418 Lopatin, N.V., Bostick, H.N., 1973, Nature of recent and fossil sedimentary organic matter (Vassojevich, N.B. ed.) Nauka, Moscow, 80-90
17. Uelsen Z 3	Ems	Kettel, D., 1981, Erdöl-Erdgas Z., 97.11. 395-404
18. Kohlhaus 1	Ruhr	Radke, M. et al., 1982, Geoch. Cosmoch. Acta 46.10. 1831-1848
19. -	NW-Germany	Bartenstein, H., Teichmüller, M., 1974, Fortschr. Geol. Rheinl. u. Westf. 24: 129-160
20. Hôd I	Pannonian	Sajgô, Cs., 1981, Adv. Org. Geoch., 1979 (Douglas, A.G., Maxwell, J.R. eds.), 103-113
21. Doboz 1	Pannonian	Laczô, I., 1984, Pers. comm.
22. Tengelic 2	W-Hungary	Halmai, J. et al., 1982, Ann. Inst. Geol. Publ. Hung. LXV. 13-113
23. Satilkovskii	Belorussia	Bogomolov, G.V. et al., 1976, Dokl. AN BSSR XX. 1. 59-61
24. Sliachovskaia Plossad	Volga	Lopatin, N.V., 1971, Izv. AN USSR Ser. Geol., 1971, 3. 95-106
25. Ust Pogosskaia 48	Volga	Lopatin, N.V., 1969, Izv. AN USSR Ser. Geol. 1971, 3:95-106
26. Verchnaia Dobrinka	Volga	
27. Pescian	Mid-Caspian	Ammosov, I.I., Eremin, I.V., 1973, C.R. 7. Congr. Int. Str. Géol. Carbonifère, 85-91 Polster, L.A. et al., 1972, Mid-Caspi oil- and gas basin, Leningrad, Nedra
28. Surgut R 51	Western Siberia	Zimin, Yu.G., 1967, Geol. i Geof. 5. 3-13 Parparova, G.M., 1966, Geol. i Geof. 7. 11-23 Neszterov, I.I. et al., 1964, Tr. VNIGRI 226

Well	Basin or location	References
29. Uzlovaia 1	N-Sakhalin	Ammosov, I.I., Utkina, A.I., 1975, Paleotemperature of zone of oil genesis (ed. Eremin, I.V.), 70-93 Olli, I.A., 1975, Organic matter and bitumen content of the sedimentary rocks of Siberia, Nauka, Novosibirsk
30. Aralsor 1	Peri-Caspian	Navrockij, O.K. et al., 1982, Geol. nefti i gaza, 1982, 4. 28-32 Bogacheva, M.I. et al., 1972, Aralsor ultradeep well. 10-106 (Vasiliev, Ju.M. ed.), Nedra, Moscow
31. Logbaba 1	Duala	Dunoyer de Segonzac, G., 1969, Mém. Carte Géol. Als. Lorr. 29 Tissot, B., Espitalié, J.,1975, Revue IFP XXX. 5. 743-777
32. DSDP Leg 50 Site 416 A	Marocco Offshore	Lancelot, Y. et al., 1980, In. Rep. DSDP L. 115-301 Cornford, C., 1980, In. Rep. DSDP L. 609-614
33. Yulleroo 1	Canning	Burne, R.V., Kanstler, A.J., 1977, J. Austral. Geol. Geophys. 2. 271-288
34. Kidson 1	Canning	Kanstler, A.J. et al., 1978, APEA J. 18. 143-156 Kanstler, A.J., 1978, Oil and Gas J., 196-205
35. Barrow Deep 1	Carnarvon	Cook, A.C., Kanstler, A.J., 1981, UN ESCAP, CCOP/SOPAC Techn. Bull. 3. 171-195
36. Bullsbrook 1	Perth	Thomas, B.M., 1979, AAPG Bull. 63.7. 1092-1107
37. Whichar Range 1	Perth	Kanstler, A.J., Cook, A.C., 1979, APEA J. 19. 94-107
38. Tirrawarra 39. Burley 1	Copper	Kanstler, A.J. et al., 1978, APEA J. 18. 143-156
40. Pelican 1	Bass	Kanstler, A.J. et al., 1978, Oil and Gas J. 196-205
41. Maui 4	Taranaki Offshore	Suggate, R.P., 1982, J. Petrol. Geol. 4.4. 377-393
42. Site V 43. Site S 44. Site N 45. Site L	Ventura Los Angeles	Bostick, N.H. et al., 1979, A Symposium in Geochemistry: Low Temperature Metamorphism of Kerogen and Clay Minerals 65-96 (Oltz, D.F. ed.)
46. Mobil Unit T-52-196	Piceance	Hood, A. et al., 1975, AAPG Bull. 59.6 986-996
47. Shell Rumberger 5	Oklahoma	----"----

Well	Basin or location	References
48.	Oklahoma	Bostick, N.H., 1973, C.R. 7. Congr. Intern. Str. Géol. Carbonifère, 183-193
49.	Gulf Coast	
50.	Gulf Coast	
51. Terrebonne Parish	Gulf Coast	
52. Cameron Parish	Gulf Coast	
53. DSDP Leg 63 Site 471	Mexico, Pacific Offshore	Yeats, R.S. et al., 1981, In. Rep. DSDP LXIII. 263-291 Rullkötter, J. et al., 1981, In. Rep. DSDP LEG 63, 819-852
54. COST GE 1	US Atlantic Offshore	Robbins, E.I., 1979, USGS Circular 800, 72-23 Scholle, P.A., 1979, USGS Circular 800, 18-32 Poag, C.W., Hall, R.E., 1979, US Geol. Surv. Circular 800, 49-63
55. COST B-3	US Atlantic Offshore	Poag, W., 1980, USGS Circular 833, 44-66 Scholle, P.A., 1980, USGS Circular 833, 13-19 Miller, R.E. et al., 1980, USGS Circular 833, 85-104
56. COST B-2	US Atlantic Offshore	Dow, W.G., 1978, AAPG Bull. 62. 1584-1606
57. Sable 4H-58	Scotian Shelf	Petrol. Geol. 30.2. 167-179
58. Sable O 47	Scotian Shelf	Powell, T.G., 1982, Bull. Can.
59. Karlsefni H-13	Labrador Offshore	Héroux, Y. et al., 1981, Can. J. Earth Sc. 18.12. 1856-1877 Kübler, B. et al., 1982, CFP Notes et Mémoires No. 17
60. Herjolf M-92	Labrador Offshore	Umpleby, D.C., 1979, Geol. Surv. of Canada, Paper 79-13. 32 Creaney, S., 1978, Geol. Surv. of Canada, Paper 78-IC, 101-103

A REACTION KINETIC APPROACH TO THE TEMPERATURE-TIME HISTORY OF SEDIMENTARY BASINS

SAJGÓ, Cs. and J. LEFLER
Laboratory of Geochemical Research of Hungarian Academy of Sciences
XI. Budaörsi út 45, H-1112 Budapest

Abstract

Three biological marker reactions have been studied in order to determine the temperature-time history of a sedimentary sequence. Two of these reactions are configurational isomerization reactions, at C-20 in a C_{29}-sterane and at C-22 in C_{31} and C_{32} hopane hydrocarbons. In the third reaction two C_{29} C-ring monoaromatic steroid hydrocarbons convert to a C_{28} triaromatic one. The progress of these reactions is different because of their different rate constants. Based on temperature and age data obtained from field measurements and on concentration measurements of reactants and products in core samples of a Pannonian borehole, we calculated the rate parameters: pre-exponential factors, enthalpies and entropies of activation.

It is obvious, that at least two different reactions are necessary to characterize the maturity of any system. The aromatization seems to be a rather complicated reaction, and we believe its use to be premature. Fortunately, two isomerizations work well and are suitable for elucidation of thermal history in different basins if the rate constants are universally valid.

The pre-exponential factors and activation enthalpies are $2.4 \cdot 10^{-3}$ s^{-1} and 91.6 kJ mol^{-1}, $3.5 \cdot 10^{-2}$ s^{-1} and 87.8 kJ mol^{-1} for isomerization of steranes and hopanes, respectively.

It was demonstrated, that by measuring the extent of the two-mentioned or other specially-chosen reactions in a lot of layers, the determination of the temperature-time history will be possible within a given temperature-time range.

Introduction

Several methods have been introduced to characterize the organic metamorphism in sediments. Recently HEROUX et al. (1979) referred to a lot of them in a comparative way. Vitrinite reflectance is a most commonly used one. A number of models were presented to describe the dependence of vitrinite reflectance both on temperature and time (duration), the most widespread ones are as follows: KARWEIL (1955); LOPATIN (1971,

1976); HOOD et al. (1975); BOSTICK (1973); BOSTICK et al. (1978); BUNTEBARTH (1978); WAPLES (1980). Other authors (a minority) neglect the role of geological time after a certain, but not too long period (NERUCHEV & PARPAROVA, 1972; AMMOSOV et al., 1977; SUGGATE, 1982; PRICE, 1983; BARKER, 1983). HOOD et al. (1975) and GRETENER & CURTIS (1982) have taken an intermediate position because they accepted the role of time only in a limited scale in oil genesis (maturation).

The basic problem in the application of vitrinite reflectance is that the variation of vitrinite reflectance is governed by the coalification of vitrinite and this process is not a genuine chemical reaction, but a complex result of different chemical reactions. The vitrinite produces various mixtures of different by-products of distinctive chemical reactions at different stages of maturation. From this it is obvious that to characterize the complex process of coalification in wide ranges may be impossible by the same rate parameters.

In connection with organic matter maturation, the following problems are also to be mentioned:

- The cited models are based on present log-temperatures at depths from which the samples have been taken, and temperature logs usually give considerable lower temperatures than the equilibrium value. BOSTICK et al. (1978) suggested that the uncertainties of the isolated values of temperature and vitrinite reflectance might be as great as 15 to 25 percent and the interpretion of geologic time might reach 40 percent uncertainty. The present temperature in the old and inactive basins may widely differ from the temperature in their active stages.

- The evaluation of thermal history is not fully solved yet taking into consideration some new tectonic models (e.g. McKENZIE, 1981; BEAUMONT et al., 1982). KELLEY et al. (1983) exploited the fission-track annealing method to determine the paleotemperature in sedimentary basins and concluded that "the thermal history of petroleum source rocks within sedimentary basins is primarily controlled not by the processes and parameters that form the basis of the tectonic stretching models, but rather by processes operating within the basin."

- The groundwater flows can cause significant thermal anomalies (e.g. ROBERTS, 1980; ZIELINSKI & BRUCHHAUSEN, 1983). Nevertheless, hydrocarbon accumulations and, moreover, oil genesis can produce thermal anomalies, too (ROBERTS, 1980; NERUCHEV et al., 1980; MOSCVIN, 1983).

The great number of the temperature-time-reflectance relationships probably arises from the erroneous estimations, measurements and interpretation of the above factors.

We will demonstrate the problem through some examples. TOTH et al. (1983) obtained a value of 0.21 ± 0.17 kJ mol^{-1} for activation energy in the range of 438 ± 140° K from modelling vitrinite values from North Sea wells. HUCK & KARWEIL (1955) determined the values of 35.16 kJ mol^{-1} and 7.16·10^3 s^{-1} for activation energy and pre-exponential factor, respectively, in the case of coalification reactions from two seams of the Ruhr district. KARWEIL (1975) calculated an activation energy of about 3.77 kJ mol^{-1} and a pre-exponential factor of 4.74·10^{14} s^{-1} for vitrinite coalification in Upper Miocene strata at US Gulf Coast, and he noticed that the Moscow lignites with these rate parameters ought, theoretically, to have been in the state of graphite.

It is also worthy to mention that LOPATIN (1971) and his followers (e.g. WAPLES, 1980) use decreasing activation energies as a function of temperature which contradicts logic and experimental observations (e.g. PETERS et al., 1977). LOPATIN (1976) realized this contradiction, and used a constant activation energy value. WAPLES's (1980) method, i.e. the conversion of TAI values (thermal alteration index) to R_o values and the correlation between the obtained R_o values and TTI values (time-temperature index of maturity), was not properly established (e.g. KATZ et al., 1982).

From this it is obvious that a lot of problems can be avoided if well-defined chemical reactions are used for maturation calculations.

Within a cooperation (McKENZIE et al., 1983) our attention was drawn by the reactions of biological marker compounds, which are suitable to study the rate laws. Later on, MACKENZIE & McKENZIE (1983) have approached the problem from the side of basin formation models while SAJGÓ et al. (1983) concentrated only on kinetics of the biological marker reactions to solve the problem of thermal history determination.

1. Biological marker reactions (Isomerizations and aromatization)

After the death of biota, some parts of their body frequently got into the water systems of earth and were buried during the geological history. During the deposition and consolidation of rocks, a significant part of the mass of dead biota decayed or was reworked by organisms, but another important part was inherited to the rocks. Numerous chemical reactions took place in this "surviving" organic mixture during deposition and early diagenesis. The greatest part of organic matter in rocks has suffered significant changes and became a part of the insoluble kerogen. A small, but very important part of the organic matter in sedimentary rocks and of oils has preserved the original carbon skeletons of the living organisms. This preservation is essentially intact so the link between biological natural products and so-called biological markers (which bear biogenic skeletons) is obvious. WHITEHEAD (1982) has recently published a review on the geochemistry of natural products and their fossil derivatives.

We have chosen two reactions (an isomerisation and an aromatisation) which involve the derivatives of sterol and another isomerisation which involves the derivatives of hopanoid precursors (e.g. bacterio-hopanetetrol). These compounds are ubiquitous in sediments.

The pathway of sterols after burial in sediments has recently been reviewed by MACKENZIE et al. (1982). As sterols of cholestane series act as rigidifiers in cell membranes, so their molecular shape is nearly flat. The sterols have 8 or 9 asymmetric atoms, but enzymes which produce these compounds seem to have been highly specific to synthesize, of the many possible configurations, the one whose shape most fitted in the organism.

In Fig. 1, we show a simplified path of sterols to steranes through sterenes. The major products of this diagenetic process are the 5α(H), 8β(H), 10β(CH$_3$), 13β(CH$_3$), 14α(H), 17α(H), 20R steranes. Either the R or the S configurations are present at C-24. During the defunctionalization the natural or biological configuration was preserved, i.e. transferred from sterols to steranes.

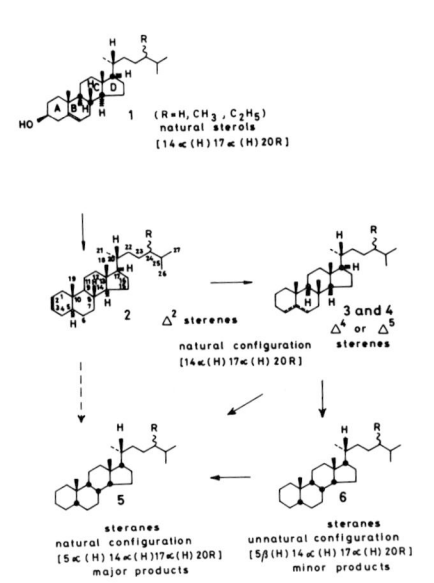

Fig. 1. Schematic pathways of natural sterols to steroid hydrocarbons during diagenesis (aromatic and rearranged ones are not included).
The hexagons and pentagons symbolize cylcohexane and cyclopentane rings. The lettering of rings and the numbering of carbon atoms are shown, in first and second structures, respectively. The conventions of thickened and dashed lines to demonstrate β and substituent and solid and open circles to demonstrate β(H) and α(H) substitution respectively, are used. α denotes a bond pointing out the page, β a bond out of the page. Arrows show the pathways of transformation. Sinuos lines mark substituent of uncertain or mixed (i.e. both α and β) stereochemistry.

In the zone of catagenesis the preserved natural skeleton suffers gradual alteration. Some isomerization reactions take place at increased temperatures (Fig. 2).

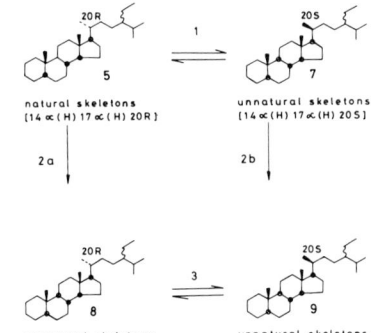

Fig. 2. The most frequent isomerizations of natural ethylcholestanes (structure 5) which take place as a function of depth (increasing temperature) in sedimentary sequences. The reaction 1 is the studied isomerization. (R = rectus i.e. right-handed; S = sinister i.e. left-handed).

The natural steranes tend to adapt the thermodynamically more stable structures (epimeric mixtures of different structures) and lose the original flatness. The enzymes synthesized all-chiral conformations with all-trans annelations (ring-junctions). These might be transformed into numerous isomers. Nevertheless, only four predominant isomers have been found in sedimentary extracts and oils (structures in Fig. 2). MULHEIRN & RYBACK (1975) have found the 5α(H), 14β(H), 17α(H), 20(R+S) steranes in geological samples but the concentrations of those resembling to 5β(H), 14α(H), 17α(H), 20(R+S) steranes are negligible.

In the natural sterane skeleton the C-8 and C-9 already have the most stable configuration and C-10 and C-13 have a CH_3-substituent whose position cannot be altered via hydrogen exchange as in the case of the other chiral centres. In the case of the C-24, MACKENZIE (1980) found the epimerization was completed at a very early stage in the Paris Basin Toarcian shales (less than 1000 m maximum burial depth). We used apolarphased gas chromatography that could not resolve these isomers so we have no genuine data about the configuration.

Subsequent to this, four chiral centers have remained for epimerization (configurational isomerization) namely at nucleus carbon atoms C-5, C-14 and C-17 and at side chain position C-20.

Indeed, in the laboratory thermal alteration studies 13 isomers have been identified (PETROV et al., 1976; SEIFERT & MOLDOWAN, 1979). Later all the possible 16 isomers have been found (PUSTIL'NIKOVA et al., 1980). Nonetheless, the conditions of isomerization were very different from the natural conditions (in a steel capsule at hydrogen pressure of 5 MPa at 570° K for 50 hr using a platinized carbon catalyst with 20 % Pt-content) much more severe. PUSTIL'NIKOVA et al. measured 45.1 % concentration for the 5α(H), 14β(H), 17β(H), (20R and 20S) isomers, 74.9 % for all the 5 isomers (8 isomers with trans-annelation of AB rings) and 73.8 % for all the 14β isomers (8 isomers with cis ring junction of CD rings). GRASS et al. (1982) have computed the thermodynamic stabilities of 13 cholestane isomers of PETROV et al. (1976) using molecular mechanics. They calculated the values of formation heat and the energy differences within the studied isomer group. Their theoretical data resemble more or less the composition of cholestane isomerate.

We have chosen the conversion of 5α(H), 14α(H), 17α(H) - 20R steranes to 5α(H), 14α(H), 17α(H) - 20S steranes (reaction 1, Fig. 2) for calculation of the reaction parameters. These compounds are the product of diagenesis (Fig. 1). The concentrations of the compounds were determined in rock extracts by computerized gas chromatograph/mass spectrometer systems mostly at Bristol. These measurements were carried out according to MACKENZIE et al. (1980). The most characteristic mass fragmentogram for the compounds is m/e-217.

The elution patterns for steranes, represented by way of m/e 217 and 218 fragmentograms (Fig. 3) are rather complex (a large number of isomers) and because of the overlap of rearranged and non-rearranged steranes the only way is to study the isomerization of C_{29} steranes (structure 1, R = C_2H_5 in Fig. 1).

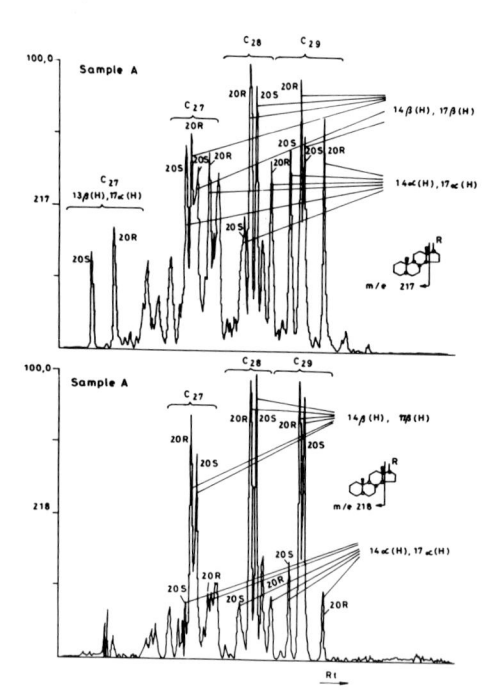

Fig. 3. Sterane mass fragmentograms of a shale extract. The m/e 217 ions are more characteristic for 14α(H) and rearranged steranes than m/e 218 ions. The m/e 218 ions are more sensitive for 14β(H) steranes than m/e 217 ones. Generally the m/e 218 fragmentograms of steranes are less complex than the m/e 217 ones of steranes.

Earlier, the m/e 217 fragmentograms were quantitated (McKENZIE et al., 1983), but later on we have found that another alkane component, with m/e 217 in its mass spectrum and a similar retention time to that of 5α(H), 14α(H), 17α(H)-20S-24-ethylcholestane was present with a slight but changing concentration in some samples. So we quantitated the m/e 218 fragmentograms which were free from this alkane instead of those of the m/e 217.

The equilibrium value was determined as a mean value of 34 samples after reaching 58 % conversion in the studied borehole. Theoretically, the two enantiomers of a chiral center need be formed in equal amounts at equilibrium since the reactants and the transition states have equal energies, so the two directions of the reaction must proceed at the same rate. If a molecule has more than one asymmetric atom (as in our case), the number of possible stereoisomers is correspondingly larger. Compounds that are stereoisomers of one another, but are not enantiomers (mirror images), are called diastereomers. Diastereomers have different physical properties, including different free energies of formation, so they need not be formed in equal amounts. Nevertheless, PETROV et al. (1976), PUSTIL'NIKOVA et al. (1980) and MACKENZIE et al. (1983) used the equilibrium value of 0.50 in calculation, but this is not correct because of the steric hindrance i.e. in the R-epimers of 17α(H) configuration, the orientation of the CH_3 at C-21 is in a gauche conformation, and in the S-epimers of the 17α(H) configurations relative to the angular CH_3 (C-18) at C-13 (see Figs. 4 and 5). So the given value of equilibrium is reasonable.

Fig. 4. An unfavorable interaction is represented between the side chain (C-20) and the C-18 methyl group in the case of 17α(H) steranes (structure I). The conversion of 17α(H) to 17β(H) minimize the torsional strain due to the interaction.

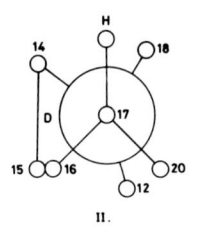

Newman type projections along
17 (13) bond. I: 5α(H), 14α(H), 17α(H) - sterane
II: 5α(H), 14β(H), 17β(H) - sterane

Fig. 5. The anti conformations (S) are more favorable in 17α(H) steranes than the gauche (R) conformations because of less interaction between the angular methyl group at C-13 and C-21 methyl group, thus the isomerization reduces the torsional strain.

The relative disposition of the 20R and 20S-steranes with 17α(H) configurations in space

At the present time, the mechanism of the reactions in this study and the contribution of the catalytic activity of the sediment are not known. The configurational isomerization (both steranes and hopanes) occurs because a hydrogen is removed from the asymmetric carbon atom at elevated temperatures probably either as a hydrogen radical (SEIFERT & MOLDOWAN, 1980) or as a hydride iron (ENSMINGER, 1977). On the basis of unpublished data, we prefer the hydrogen radical mechanism to hydride ion one. ABBOTT et al. (1984,1985) found that the isomerization took place under free radical conditions at the chiral centres in pristane. The radical intermediate (or carbonium ion, too) is almost planar for acyclic carbons and has a close to equal probability of regaining a hydrogen radical (or a hydride ion) on either side. In such a way, the biologically inherited sterol configuration at C-20 converts to a near equal mixture of the R and S epimers.

Some catalytic effect of host rock should be supposed, as SAJGÓ et al. (1983) have found different extents of isomerizations and aromatization in the "open" and "closed" pores of the same sample. The greater extent of these reactions in closed pores (i.e. 27 % for steranes at C-20, 33 % for hopanes at C-22 and 38 % triaromatization) than those in open pores (i.e. 5 %, 20 % and 9 %, respectively) suggest that the pore structure and chemistry of the pore linings have an important role at such reactions in biomarker chemistry. The rates of these reactions may control the yields through the presumed necessary adsorption-desorption processes. Nevertheless, if the ratio of the open and closed pores does not vary in a sequence or the sampling is representative, this catalytic effect does not disturb the evaluation.

The other isomerization we considered is the conversion of the 17α(H), 21β(H)-22R-homohopane and bishomohopane to a mixture of 22R and 22S hopanes. The hopanes and their precursors belong to natural pentacyclic triterpenoids. The members of the hopane family (or hopanoids) found in lower order organisms (procaryotes) bear a five-membered E-ring in contrast to the equivalent six-membered ring of more modern triterpenoids in eucaryotes. OURISSON et al. (1979) provided a detailed review about the family. They found that the extended hopanes (over C_{30}-members) could be derived from the bacteriohopanetetrol (Fig. 6). They also supposed that the hopanoids serve as "rigidifiers" for cell wall membranes in procaryotes-like sterols in more devel-

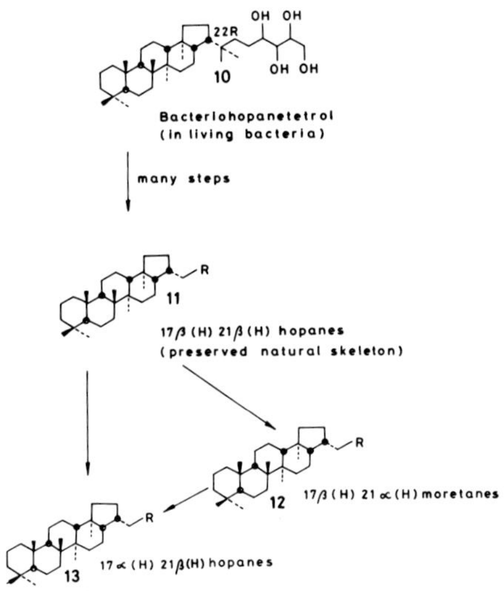

Fig. 6. The simplified pathways of 17α(H), 21β(H) hopanes from bacteriohopanetetrol

oped biota. RULLKÖTTER & PHILP (1981) found hopanes up to C_{40} in a bitumen which means that other biological precursors would also be required for higher hopanoids.

In the case of the most stable hopanoid hydrocarbons 17α(H), 21β(H)-hopanes a configurational isomerization occurs at C-22 (Fig. 7) resembling that in steranes at C-20. This epimerisation can be observed in the C_{31} - C_{35} hopanes, as revealed by

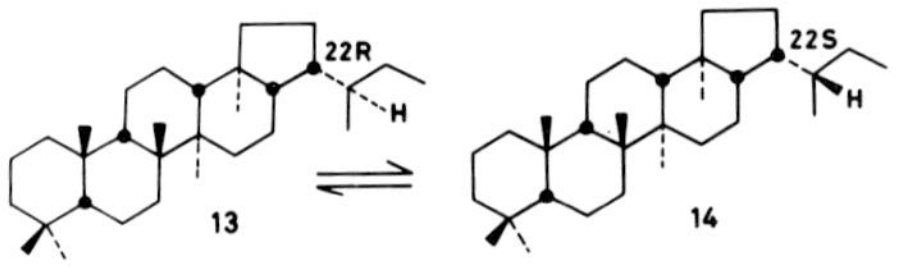

Fig. 7. The configurational isomerization of homohopane at C-22

the m/e 191 fragmentograms for the GC - MS analyses of separated alkane fractions. We averaged C_{31} and C_{32} isomers to decrease errors. Gammacerane (a pentacyclic triterpane with six-membered E-ring) was not detected in the samples studied, for it can overlap with the 22R-C_{31} isomer (e.g. SHI JIYANG et al., 1982). The equilibrium value of 57 % was obtained as the mean value of 26 samples deeper than 3100 m.

The reaction mechanism of the above isomerization as well as at C-20 in steranes has not been proved yet, but the two epimerizations must have the same mechanism because of the close structural similarity between the compound families (e.g. ENSMINGER et al., 1978). Nevertheless, hopane isomerizations occur at a faster rate and this suggests some dissimilitude, too (e.g. hopanoids contain methyl groups on both sides of the nucleus, whereas methyl groups attached to sterane molecules are directed only towards one side).

It is worthy to mention that COSTA NETO (1983) suggested an alternative model for the configurational isomerization of hydrocarbon chiral centres in sediments. He believes it is theoretically possible, for direct protonation of chiral carbons, to form positively-charged pentavalent carbon atoms, allowing the asymmetric carbon to cross through this face and allowing the other enantiomer to be formed. The proposed model suggests that the greater the pressure in the sediments, the greater will be the extent of isomerization.

The third reaction we studied is the aromatization of C-ring monoaromatic steroid hydrocarbons to ABC-ring triaromatic steroid hydrocarbons with a nuclear methyl group via diaromatics (Fig. 8). During the transformation the loss of the nuclear methyl group (C-19) on the A/B-ring juncture and of seven protons takes place probably in several steps.

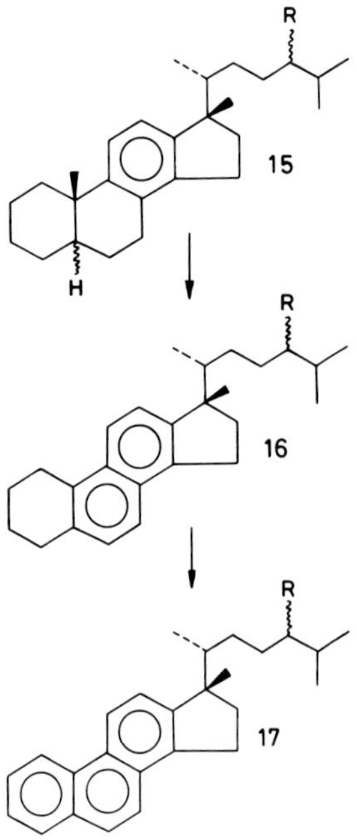

Fig. 8. The simplified conversion of C-ring monoaromatics (four isomers in the case of the C_{27}-C_{29}; structure 15) to ABC-ring triaromatics (two isomers in the case of the C_{26-28}; structure 18) via less stable diaromatics (structure 16). Inscribed circles in hexagons denote aromatic rings.

Ring C monoaromatic steroid hydrocarbons occur in sediments and petroleum, and seem to be generated just before steranes (LUDWIG et al., 1981a; MACKENZIE et al., 1982; SEIFERT et al., 1983). They may derive from diagenetic dehydration of sterols via the rearrangement of the stera-3.5-dienes (MACKENZIE et al., 1982). The C-ring monoaromatic steroids were produced in laboratory thermal experiments (ZUBENKO et al., 1980; SEIFERT et al., 1983). Their structure were proved by LUDWIG et al. (1981b), ZUBENKO et al. (1981), and SEIFERT et al. (1983), i.e. C-ring monoaromatics in geological samples have methyl groups at C-10 and at C-17 and the stereochemistry of the known ring C monoaromatic geosteroids is 10β(CH_3), 17β(CH_3) with three chiral centres (C-5, C-20 and C-24; see Fig. 8 structure 15). Both forms exist at each centre, but the mixture of isomers at C-24 is not resolved by the apolar gas chromatography columns which are generally used. Thus, there are four C-ring monoaromatic steroid hydrocarbons isomeric at C-5 and C-20 (not including C-24) giving four peaks per carbon number (C_{27}-C_{29}). We studied only C_{29} C-ring monoaromatics with 20R stereochemistry and C_{10} side chains (four compounds isomeric at C-5 and C-24, but two peaks). The key ion i.e. the base peak in the mass spectra of the ring C aromatic steroid hydrocarbons is m/z 253 (MACKENZIE et al., 1981).

Within the zone of late diagenesis and catagenesis the triaromatic steroid hydrocarbons arise from monoaromatic ones. There are four triaromatic series whose nuclei have none, or up to three methyl substituents (MACKENZIE et al., 1981a; LUDWIG et al., 1981). We concentrate only on those which have one methyl group. This series is widespread in sediments and petroleums, and they are the most abundant amid triaromatic ones. Their structure has been proved by synthesis and comparison with the natural products (LUDWIG et al., 1981).

MACKENZIE et al. (1981b), SEIFERT et al. (1983) and ABBOTT et al. (1984,1985) converted C-ring monoaromatics to ABC-ring, triaromatic steroid hydrocarbons in laboratory heating experiments. ABBOTT et al. (1984) brought about the above aromatization under free radical conditions. The ABC-ring triaromatics have three chiral centres (C-17, C-20 and C-24) because the chiral centre at C-5 has been lost. The position of the methyl group at the quaternary C-17 is fixed as 17β(CH_3) and we measured only the 20R stereochemistry of C_{28} triaromatic steroid hydrocarbons, thus the unresolved C-24 isomers remained (Fig. 8, structure 17) giving only one peak in the m/e 231 fragmentogram. The BC-ring diaromatic compounds (Fig. 8, structure 16) have also been recognized by SCHAEFLÉ (1979) and MACKENZIE et al. (1981a) but only in small concentrations. MACKENZIE et al. (1982) suggested that the aromatization of ring B in C-ring monoaromatics had to be followed almost immediately by the aromatization of ring A.

From the above, the conversion of the proposed reaction (Fig. 8) can be characterized by relative concentrations represented by three GC-MS-peaks. The GC-MS measurements were carried out according to MACKENZIE et al. (1981a) and SHI JIYANG et al. (1982) from m/e 253 for monoaromatics and from m/e 231 for triaromatics.

Finally, we allude to the problem of the oil formation, i.e. during catagenetic hydrocarbon generation biological markers are released from kerogen and asphaltene in considerable amounts. It is not clear yet whether the extents of isomerizations and aromatizations are the same in the free and bound steroids and hopanoids. Since if the reactions studied are faster in free compounds than in the bound ones inside the kerogen matrix, then the released (secondary) steroids and terpanes would modify the extent of reactions obtaining an unrealistic conversion. In other words, it is straightforward to suppose that the reaction rate in the free state and in solution is different from that in the solid bound state. GALLEGOS (1975), SEIFERT (1978), SEIFERT & MOLDOWAN (1980), RUBINSTEIN et al. (1979), CURIALE et al. (1983) and RULL-KÖTTER et al. (1984) studied the biological marker products of the thermal breakdown of kerogen, asphaltene and solid bitumens in laboratory pyrolysis. Their results did not give a definite answer to the above question. The differences emphasize the care that must be taken within the zone of oil generation. In our case, fortunately, the reactions studied reached their equilibria or went to completion before the threshold of the ongoing, intense oil generation (SAJGÓ, 1980). Thus, we could neglect this problem hereafter.

2. The determination of rate parameters of the biological marker reactions
2.1 Geological setting of the borehole Hód-I

The core samples investigated come from the Hód-I borehole in the southeastern part of Hungary. The region in question is inside the Carpathian arc, which is an integral part of the Alpine-Himalayan mountain belt, and it is called the Carpathian basin. Inside the Carpathian basin there is the Pannonian basin surrounded by peripheral basins. The formation of the basins was hypothesized by several authors (e.g. STEGENA, 1967; SZÁDECZKY-KARDOSS, 1971; SCLATER et al., 1980; HORVÁTH & ROYDEN, 1981). A recent volume of the Earth Evolution Sciences (1981/3-4) was devoted to the evolution of the Carpathian basin.

The Hód-I was drilled in the Makó trough nearly in central position, encircled by a depression. The area under consideration is the southeastern part of the Great Hungarian Plain, formed during the Neogene as part of the Pannonian basin. The borehole, which is 5842 m deep, has not penetrated formations older than Badenian, Middle Miocene (about 15 Ma). MUCSI & RÉVÉSZ (1976) suggested that the last sedimentary cycle (from the Helvetian up to present) was uniform and free of intermediate greater regressions.

In Hód-I it is surprising that although the strata seem to be continuous no Sarmatian faunae were found between the Lower Pannonian and Badenian sediments. The total sequence represents a marine regression: the Badenian sediments are marine whilst the Lower Pannonian varies between deltaic and littoral, and after gradual transition during the Upper Pannonian, lacustrine-fluviatile sediments have deposited.

The sedimentary sequence of the Hód-I was introduced earlier (SAJGÓ, 1980). The problems of oil generation and maturation were also discussed in the case of the Hód-I borehole previously (SAJGÓ, 1980; SAJGÓ et al., 1983).

Table 1 presents the age and temperature data which were taken from SAJGÓ (1980). Calculating age data, the sedimentation rates within the Upper Pannonian and within the Lower Pannonian stages were considered permanent, 500 m/Ma and 431 m/Ma, respectively.

The present temperature at depth, from which the samples were taken, was accepted as the maximum temperature of samples. This assumption is very probable because of the fast continuous subsidence of the young sediments up to the present. The Pannonian basin is relatively hot and has a high heat flow. The heat flow has a value of about 100 mW/m² in the Great Hungarian Plain (HORVÁTH et al., 1981; STEGENA et al., 1981). VÖLGYI (1977) reported a value of 55° C/km as an average temperature gradient for the Great Hungarian Plain determined from 1400 data. This average value is higher than that in the Hód-I (38° C/km).

Only cores were used in this study. Their surfaces were cleaned several times and checked strictly under u.v. light, and in the case of contamination their outer parts (about 1 cm in thickness) were cut off before grinding. The samples are clay marls and siltstones with carbonate-contents between 25 - 40 percent. The C_{org}-content varies between 0.25 and 0.54 percent, with an 0.36 percent average. The chloroform extract varies between 130 and 390 ppm and the average is 250 ppm. The chloroform extract/C_{org} (mg/g) ratio shows a variation between 41 - 98 with an average of 66. The uppermost sample has 1.32 %, 500 ppm and 38 values, respectively. The bitumens are autochthonous according to i.r. analysis. The kerogen varies between type II (predominant) and type III.

Table 1. Geological and geochemical data

Sample No.	Isomerization		Aromatization	Temperature	Age
	Sterane x	Hopane y	z	°C	(Ma)
H-I/ 1	0.05	0.15	0.18	89	4.90
H-I/ 2	0.08	0.24	0.24	98	5.38
H-I/ 3	0.14	0.34	0.44	101	5.58
H-I/ 4	0.18	0.42	0.92	106	5.85
H-I/ 7	0.25	0.50	0.95	113	6.33
H-I/ 8	0.28	0.53	0.97	116	6.47
H-I/ 9	0.41	0.57	0.99	121	6.82
H-I/10	0.46	0.57	1.00	123	6.93
H-I/12	0.51	0.58	1.00	129	7.27
H-I/14	0.55	0.57	1.00	135	7.67
H-I/15	0.57	0.57	1.00	137	7.82

2.2 The physico-chemical fundamentals of the determination of rate parameters

The rate of a chemical process depends on the rate of the chemical reaction and the rate of the mass transfer. The contribution of the mass transfer to the rate of process can be neglected in the case of these biological marker reactions studied. We are allowed to proceed so because the reactants and products do not leave the system, or if they do so slightly, the reactants and products are involved at the same level. These considerations are not fully true for aromatizations because:

i) The lost hydrogen atoms and methyl group can obviously leave the system much easier than the starting monoaromatic HC and the forming triaromatic HC.

ii) The monoaromatic steranes are less prone to migration than triaromatic ones according to the polarity and molecular shape differences between the two families.

iii) The aromatization may be considered to proceed to completion, unlike the isomerizations which go to equilibrium mixtures and this practically means that no mass transfer takes place from the viewpoint of the reactions.

In chemical kinetics the reaction rate is defined as the rate of concentration changes of the reactants or the products which are, in our case, R-isomers or monoaromatics and S-isomers or triaromatics, respectively:

$$r = \frac{dC}{d\tau}$$ where C is concentration and τ is the time.

The mechanism of reactions governs the order of reactions. The reaction order can be written by the kinetic equation. A first-order reaction obeys the equation

(1) $$r = \frac{dC_A}{d\tau} = k\, C_A$$

where C represents the concentration in arbitrary units and k is rate constant in s^{-1}.

The second-order reactions can be written as

(2) $$r = k\, C_A \cdot C_B \quad \text{or} \quad r = k\, C_A^2$$

There are reactions which can be described - more or less formally - by reactions zero-, fraction- and higher-orders, too. In complicated cases we usually have complex reactions with several sequential and/or parallel steps between reactants and final products. Theoretical considerations show that the isomerizations studied are unimolecular first-order reactions. McKENZIE et al. (1983), and MACKENZIE & McKENZIE

(1983) have found it so, too. In the cases of both reactions it is necessary for the transforming molecule to borrow an energy surplus to surmount an energy barrier. This barrier is called the activation energy (E_a) or the enthalpy of activation (ΔH^*), which are not necessarily equal.

The mechanism of isomerization is not known yet, but we can hypothesize as follows:

The reactant molecules are continually colliding with the molecules of the host-rock and exchanging kinetic energy. Only a few will obtain sufficient energy for epimerization. The surplus energy in excited species will be redistributed among all the available individual vibration-rotation states. The disposal of additional energy takes place according to Boltzmann distribution among all the states and so a very few molecules have sufficient energy at C-20 to overcome the barrier to configurational isomerization.

The isomerizations in these molecules take place within a vibrational period (ca. 10^{-13} s) and this means that the rearrangement proceeds during the collision. Consequently, the reaction is strictly bimolecular, but because the quality of the other colliding species has not any particular role we can consider the process as unimolecular without further ado.

As the Boltzmann distributions govern the number of the reacting molecules only a higher temperature will shift the entire distribution to higher energies, i.e. the ratio of reacting molecules to all the molecules is a constant which depends only on temperature. As the quantity of the collision excited reacting molecules is constant so we can consider it as an equilibrium value of the reaction and consequently the reaction rate can be written as a first-order kinetic equation.

Microscopically, the process of aromatization is much more complex than that of isomerizations. The conversion of a monoaromatic molecule to a triaromatic one, via the loss of seven hydrogen atoms and a methyl group, consists, necessarily, of a lot of elementary steps and the localized bonds become the parts of the 14 delocalized bonds. Despite its complexity, the aromatization can be characterized by a first-order rate low. This is probably because one of the sequential steps, the slowest (rate-determining), obeys a first-order law. The chance of a catalytic effect of environment (host rock) is greater in the case of aromatization than in the case of simpler isomerization. Another important difference is that aromatization is irreversible and the isomerization is reversible in the cases studied.

The R and S structures of steranes and hopanes at C-20 and C-22, respectively, are nearly equally stable, therefore in the final equilibrium mixture they are present in nearly equal amounts. This means that the conversion of the biologically preferred R isomers to S isomers starts after the decay of organisms.

We shall treat relationships between the GC-MS measurements of the concentrations and the rate parameters (activation enthalpy, pre-exponential factor and standard entropy change) for the more common reversible reactions and the relationships for the irreversible reaction will be deduced from the above as marginal cases.

Table I consists of the present temperature, geological age and conversion (x, y and z) data of the studied core samples. The meanings of conversions are

$$(3) \quad x = \frac{20S}{20S+20R}; \quad y = \frac{22S}{22S+22R}; \quad z = \frac{T}{M+T}$$

where 20S, 20R, 22S, 22R, T and M stand for the concentrations of 20S, 20R, 22S, 22R, T(riaromatic) and M(onoaromatic) hydrocarbons in the samples, respectively.

If we write the scheme of a general reaction as

$$(4) \quad A \rightleftharpoons B$$

the rate of the reaction can be written for forming B as

$$(5a) \quad \frac{dC_B}{d\tau} = k_A \cdot C_A - k_B \cdot C_B$$

or for forming A as

$$(5b) \quad \frac{dC_A}{d\tau} = k_B \cdot C_B - k_A \cdot C_A$$

At equilibrium the rates of forward and reverse processes are equal, consequently

$$(6) \quad k_B \, C_B^+ = k_A \, C_A^+$$

where C_A^+ and C_B^+ are the concentrations of A and B at equilibrium.

We can rewrite equation (6) utilizing the equilibrium constant of reaction (K) as

$$(7) \quad \frac{k_A}{k_B} = \frac{C_B^+}{C_A^+} = K$$

If we introduce a ratio, calculated from concentration measurements

$$(8) \quad \alpha = \frac{C_A}{C_A + C_B}$$

then we can write the solution of the differential equation or rate law (from equations 5b, 7 and 8) as

$$(9) \quad -\frac{1}{\beta} \ln(1-\beta\alpha) = \int_0^\tau k_A \, d\tau$$

where k_A is the forward rate constant and

$$(10) \quad \beta = \frac{K+1}{K}$$

The irreversibility of the aromatization means that $k_A \gg k_B$, therefore $\beta = 1$. In the studied isomerizations the rate constants are nearly equal ($k_A \approx k_B$) and β subsequently is close to 2. There are numerous ways to calculate the equilibrium constant. One of them is based on laboratory heating experiments. PUSTIL'NIKOVA et al. (1980) applied pressurized heating experiments to study the isomerization of cholestane. The conditions have been referred to earlier. We must bear in mind the difficulties of this method. The first question is whether the mixture is really at equilibrium. To reach equilibrium is a long process, and in geological situations it may require some million years. To speed up the process, PUSTIL'NIKOVA and her co-workers have used a temperature about 150 - 200° C higher than natural conditions. In replicate simulations, the equilibrium constant for 20S/20R, $5\alpha(H)$, $14\alpha(H)$, $17\alpha(H)$ cholestanes has been found to be 0.82 and 0.98. Considering the dependence of the equilibrium constant on temperature we must be very careful if we apply the values determined at high temperatures to natural conditions. From a kinetical viewpoint, the presence of pressurized H_2 and Pt-catalyst is crucial, because it can modify the mechanism of the reaction and therefore the rate parameters may change drastically.

Van GRASS et al. (1982) have computed the thermodynamic stabilities of 13 cholestane isomers which were found in Petrov's isomerate (SEIFERT & MOLDOWAN, 1979) using a molecular model based on empirical energy functions. The equilibrium constant can be derived from the obtained composition of equilibrium mixtures of isomers. In this case the inverse of the above-mentioned equilibrium constant ($\frac{1}{K}$) was found to be 0.817 at 298° K and 0.845 at 573° K. The reliability of such model computations depends on the reality of the established models and on the consideration of all the possible interactions.

We have chosen a third way for the calculation of equilibrium constants, namely, starting from measurements on geological samples. The concentration ratio in geological samples of the two isomers studied can be considered constant for a duration and over a certain temperature. For example, in the case of sterane isomerization in a sample whose age is 7.82 Ma and has a temperature of 410° K this ratio has reached the equilibrium value, whereas in older samples the ratio scatters irregularly. The scale of fluctuation in the decisive part of samples does not exceed the measurement errors (MACKENZIE, 1980; RULLKÖTTER et al., 1984) and this means that the x, y and z values (Table 1 and equation 3) can be determined to an accuracy of ± 0.04. The reciprocal equilibrium constant was calculated from the average of concentration values and, for example, it was found 0.725 for the above mentioned sterane isomerization, so in our case it was found less than that in van GRASS et al. (1982) (0.817 at 298° K and 0.845 at 573° K).

The integration of the rate-law equation sets some problems, since the rate constant behind the integral symbol has a temperature dependence, which follows the classic equation proposed by Arrhenius in 1889:

(11) $\quad k = A \exp [- \frac{\Delta H^*}{RT}]$

where A is the pre-exponential factor; ΔH^* is the activation enthalpy, i.e. the height of the potential barrier between the reactants and products or, in other words, the additional energy of the activated complex with respect to the initial state of reactants; R is the universal gas constant and T is the absolute temperature.

According to more detailed molecular theories the pre-exponential factor has a temperature dependence which is however much less than that of the rate constant:

(12) $\quad A = X \cdot \frac{\bar{k} T}{h} \exp [\frac{\Delta S^*}{R}] = A' T$

where X is the transmission coefficient or the proportionality constant, $X \leqslant 1$ but its value is usually 1 - its deviation from one is proportional to the reduced probability of reformation of the reactants from the activated complex - $\bar{k}$ is the Boltzmann constant; h is Planck's constant - the ratio $\frac{\bar{k} T}{h}$ is termed the fundamental frequency and its value at room temperature (T = 300° K) is $6 \cdot 10^{12}$ s^{-1}, thus it is not far from the value of collision frequency (10^{13} s^{-1}) ; ΔS^* is the activation entropy of reaction and it is an indicator of the configuration of the activated complex. ΔS^* is usually negative because $A < \frac{\bar{k} T}{h}$. This decrease in entropy is a consequence of the loss of translational and rotational freedom when reactants are combined to form the activated complex.

In laboratory simulation experiments we can ensure the isothermal conditions - both in space and time - therefore we can eliminate the thermal dependence of the rate constant during the solution of the rate law equation. In such a case the integration becomes a simple multiplication. Unfortunately, the variation of temperature with time and place is characteristic of geochemical processes. As the reaction rate is temperature-dependent through the exponential dependence of the rate constant on temperature, in cases of the same duration the different temperatures can produce considerably dissimilar conversions. This feature of reactions can be exploited to elucidate the thermal history within a subsiding basin with help of the analyses of reactants and products of biomarker reactions. If we know the rate parameters (A, ΔH^* and ΔS^*) of a reaction we can calculate the extent of conversion in the case of any thermal history - $T = f(h,\tau)$ - or inversely, the thermal history can be reconstructed by the extent of conversion in a series of samples. For this reason we attached great importance to determine the genuine rate parameters of the studied reactions. Knowing the temperature and age data of the samples (Table 1), since the present temperature is maximum in the Pannonian basin and the age of the samples in the quickly subsiding basin can be determined accurately, and having the conversion data of the compounds of the three reactions (also Table 1), we can give the rate parameters of the isomerizations of the sterane and the hopane, and the aromatizations of the monoaromatic steroid. These parameters may also be exploitable to describe the thermal history of other basins.

Some authors tried to utilize the fact that the conversion depends on temperature exponentially and on time linearly during the solution of the integral equation of the rate law. For example, HOOD et al. (1975) suggested that the effective heating time (τ_{eff}), the time which the sample spent within 15° C of its maximum temperature, could be used to characterize the transformation of the vitrinite instead of the whole age of the sample. Others (NERUCHEV & PARPAROVA, 1972; SUGGATE, 1982) suggested that about one million years are enough for organic geochemical reactions (e.g. coalification) to reach equilibrium. These models (e.g. HOOD et al., 1975) using the present temperature and the variations of the geothermal gradient with time can be very elusive if they are applied. The organic geochemical reactions go at any case in a given direction from the viewpoint of final products. This may be some sort of equilibrium mixture of the reactants and the products, or, in the case of an irreversible reaction, the final products. On the other hand, the rate of each chemical reaction practically increases with rising temperature. So, on the basis of the above it is obvious that a conversion having proceeded at a higher temperature does not go back after a substantial temperature decrease. Strictly speaking, however, in the case of reversible reactions such a reverse process can be imagined if the lower temperature is favorable to the starting compounds in consequence of temperature dependence. But in the case of the isomerizations chosen, the temperature dependence is so little that its influence can be neglected knowing the temperature variation of different basins.

When the case is simpler, e.g. a constant rate of subsidence with a constant geothermal gradient or with a monotonic increase of heat flow density through a longer period, the influences of the temperature and the age on the rate parameters are controlled by the extent of the activation energy of the reaction. Mathematically, the operation is correct only if we take into consideration the temperature dependence of the rate constant and the time dependence of temperature during the course of integration (from equations 9 and 12):

$$(13) \quad -\frac{1}{\beta} \ln(1-\beta\alpha) = \int_0^\tau x \frac{\bar{k} T}{h} \exp\left[\frac{\Delta S^*}{R}\right] \exp\left[\frac{-\Delta H^*}{RT}\right] d\tau$$

We can generally describe the temperature of a subsiding and transforming sediment sample as a function of time and depth during the course of basin formation as

$$(14) \quad T = f(\tau)$$

and the inverse function:

$$(15) \quad \tau = g(T)$$

Inserting these functions into the reaction integral we get:

$$(16a) \quad -\frac{1}{\beta} \ln(1-\beta\alpha) = \frac{x \bar{k}}{h} \exp\left[\frac{\Delta S^*}{R}\right] \int_0^\tau f(\tau) \exp\left[\frac{-\Delta H^*}{Rf(\tau)}\right] d\tau$$

or

$$(16b) \quad -\frac{1}{\beta} \ln(1-\beta\alpha) = \frac{x \bar{k}}{h} \exp\left[\frac{\Delta S^*}{R}\right] \int_{\tau_0}^\tau T\, g'(T) \exp\left[\frac{\Delta H^*}{RT}\right] d\tau$$

where

$$g' \text{ is } \frac{g(T)}{\tau}.$$

Henceforth the above means of solving the reaction integral will be called the method of absolute time (or in abbreviated form: MAT). This method is definitely the best to determine the rate parameters (A, ΔH^* and ΔS^*) from the measured extents of conversions and from knowing the temperature function, $T = f(\tau)$, or backwards to reconstruct the temperature history on the basis of the known rate parameters and the measured extents of conversions. However, the application of the method has some difficulties:

i) The primitive function cannot usually be written in a closed from (or in an easy tractable form) in the case of the $T = f(\tau)$ formula; the thermal history of the Pannonian basin was regarded as a simple case:

(17) $\quad T = T_0 + a\tau$

and the primitive function may be given as:

(18) $\quad \int Tg'(T) \exp[\frac{-\Delta H^*}{RT}] \, dT = a^{-1} \left(\frac{T^2}{2} - (\frac{\Delta H^*}{R})T - \frac{1}{2!}(\frac{\Delta H^*}{R})^2 \ln T + \right.$
$\left. + (\frac{\Delta H^*}{R})^2 \left\{ \sum_{n=1}^{\infty} (-1)^{n+1} \frac{1}{n(n+2)!} (\frac{\Delta H^*}{R})^n T^{-n} \right\} \right)$

Unfortunately this series converges rather slowly so its application is not convenient.

ii) There is no exact expressions for ΔH^* in an explicit form from the reaction integral, thus the knowledge of its approximate value is desirable.

To solve the first problem we can use numerical integration methods, e.g. Simpson's approximation or Romberg integration which work relatively quickly and with an accuracy according to choice.

We can solve the second problem in two steps. First we determine ΔH^* approximately in some way, and then we insert the approximate ΔH^* into the reaction integral to compute the value of ΔS^*. Having two parameters we can calculate the rate constants belonging to present temperature of the samples. From these values we calculate ΔH^* again and utilizing its new value we get a new set of ΔS^*. We proceed with the iteration up to obtaining the least dispersion of the series of ΔS^*.

The enthalpy of activation is obtainable by the differential method in the simplest way. The method was based on the following train of thought. As we have shown (in equations 12 and 13):

(19) $\quad -\frac{1}{\beta} \ln(1-\beta\alpha) = A' \int_0^\tau T \exp[\frac{-\Delta H^*}{RT}] \, d\tau$

and carrying out differentiation on both sides of the equation:

(20) $\quad \frac{1}{1-\alpha\beta} \frac{d\alpha}{d\tau} = A'T \exp[\frac{-\Delta H^*}{RT}] = k$

Substituting the finite $\frac{\Delta\alpha}{\Delta\tau}$ for $\frac{d\alpha}{d\tau}$ we have determined ΔH^* and A' (and thereby ΔS^*) approximately. The accuracy of determination for parameters was controlled by the number of the available survey data in this case apart from the accuracy of measurement.

We can assess the rate parameters with the help of the effective heating time (EHT) method.

Applying this method the reaction integral can be given as

(21) $\quad -\frac{1}{\beta} \ln(1-\beta\alpha) = k \Delta\tau$

where $\Delta\tau$ is the effective heating time and subsequently it may be written (from equation 19) as

(22) $\quad -\frac{1}{\beta} \ln(1-\beta\alpha) = A'T^o \exp[\frac{-\Delta H^*}{RT^o}] \Delta\tau$

where T^o is the present temperature. So we can rewrite as

(23) $\quad \ln[-\frac{1}{\beta T^o} \ln(1-\beta\alpha)] = \ln A' + \ln \Delta\tau - \frac{\Delta H^*}{R} \cdot \frac{1}{T^o} = x$

This expression determines the equation of a straight line in a coordinate system of x vs. $\frac{1}{T^o}$. ΔH^* can be obtained from the slope of line and the intercept yields A' which then gives ΔS^*.

The error of the EHT method can be assessed as follows:

(24) $\quad \int_0^\tau e^{-\frac{\Delta H^*}{RT}} d\tau = \int_0^{\tau-\Delta\tau} e^{-\frac{\Delta H^*}{RT}} d\tau + \int_{\tau-\Delta\tau}^\tau e^{-\frac{\Delta H^*}{RT}} d\tau = I$

on the basis of the mean value theorem of integral calculus:

(25) $\quad I = (\tau-\Delta\tau)e^{-\frac{\Delta H^*}{RT'}} + \Delta\tau\, e^{-\frac{\Delta H^*}{RT''}}$

where T' is the temperature in the time interval $(0, \tau-\Delta\tau)$, and T'' is the temperature in the interval $(\tau-\Delta\tau, \tau)$. The first term on the right-hand side is nearly the same as the expression on the right-hand side for the method of effective heating time (equation 22). The only difference is that in this case the present temperature has been replaced by T'' which is different. If the interval of effective heating time is not too long, then

(26) $\quad T'' = T^0 - f(\tau) \Delta\tau \xi$

where using $0 < \xi < 1$, we can get

(27) $\quad \Delta\tau \exp[\frac{-\Delta H^*}{RT''}] = \Delta\tau [\exp(\frac{-\Delta H^*}{RT})]^{1/J}$

where

(28) $\quad J = 1 - \frac{f'(\tau)\Delta\tau \xi}{T^0}$

The first source of error was established by considering $J = 1$ in the method of effective heating time. This did not cause a considerable deviation because $J = 0.95$ in the Pannonian basin. The first term of the sum is more than zero, and this fact yielded the second source of error. The more considerable the error, the higher is T'. Its upper limit is $T' = T^0 - f'(\tau)\Delta\tau$ if the temperature is continuously rising in the samples. Thus we can substitute the first term of the expression for the following formula:

(29) $\quad (\tau-\Delta\tau) \exp(\frac{-\Delta H^*}{RT'}) \leq \frac{\tau-\Delta\tau}{\tau} [\exp(\frac{-\Delta H^*}{RT^0})]^{1/J'}$

where

(30) $\quad J' = 1 - \frac{f'(\tau)\Delta\tau}{T^0}$

If the heat flow had a maximum in the past, T' should be replaced by the maximum temperature as an upper limit and the increase in error would be substantial.

To turn back the train of thought prior to assessing the error of the effective heating time method, it can be obtained by:

(31) $\quad -\frac{1}{\beta\tau} \ln(1-\beta\alpha) = A [\exp(\frac{-\Delta H^*}{RT^0})]^{1/J'}$

where $J' = 1 - \frac{f'(\tau)\Delta\tau}{T^0}$ from the assessment of error, and we can write this as

(32) $\quad -\frac{1}{\beta} \ln(1-\alpha\beta) = k \tau_g$

where τ_g is the whole geological age, i.e. the time since deposition.

This formula shows that it is necessary to consider the whole time span when we calculate the rate parameters of the conversions. From the above expression, of course, we can estimate the rate parameters as we have done in the case of the effective heating time method. We shall show using this assumption in the samples of Pannonian basin that the determination of the rate parameters for the isomerization of steranes and hopanes and for the aromatization of the C-ring monoaromatic steroids is not more unreliable than it is using the effective heating time method. The empiric fact that the values of rate parameters within precision of measurements are independent of the used duration of reaction in the course of the calculation, has no universal validity on the reactions in question for the Pannonian basin. This observation is due to three factors: the relative high enthalpy of activation, the rapidly and monotonically increasing temperature in the subsiding basin, and the inaccuracy of measurements.

2.3 The calculation of the rate parameters for the reactions studied

1) Isomerization of sterane at C-20

The equilibrium constant for sterane isomerization from geological samples gives $K = 1.38$. Subsequently β is 1.724. Fig. 9 shows how the rate constant calculated with the method of effective heating time depends on temperature. The effective

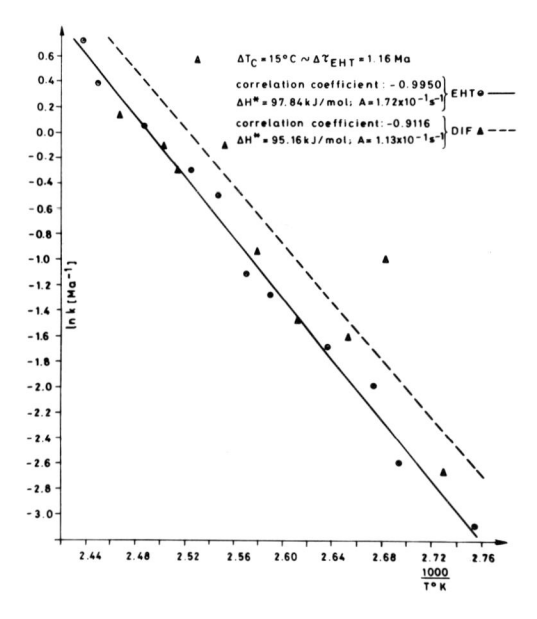

Fig. 9. Arrhenius plot for sterane isomerization at C-20 in Pannonian sequence of Hód-I, obtained from EHT (effective heating time) and differential methods (DIF). The latter shows considerable dispersion caused by the insufficient available survey data.

heating time value of 1.16 Ma was applied on the basis of geology (SAJGÓ, 1980). From the slope and intercept of the straight line we obtained the enthalpy of activation, $\Delta H^* = 97.84$ kJ mol^{-1}, and the pre-exponential factor, $A = 1.13 \cdot 10^{-1}$ s^{-1}. The entropy of activation calculated from the value of A gives $\Delta S^* = -264.9$ J mol^{-1} °K^{-1}.

The Arrhenius plot (lnk - 1/T) based on the differential method is also shown in Fig. 9.

The calculation was repeated for all the samples using the whole geologic age not only EHT of samples (Fig. 10). The value of activation enthalpy shows a good agreement with the value estimated by the effective heating time method, $\Delta H^* = 90.14$ kJ mol^{-1}. The value of A of $7.986 \cdot 10^{-4}$ s^{-1} is, however, much less. Subsequently, the entropy of activation is less (more negative value), $\Delta S^* = -314.3$ J mol^{-1} °K^{-1} than earlier.

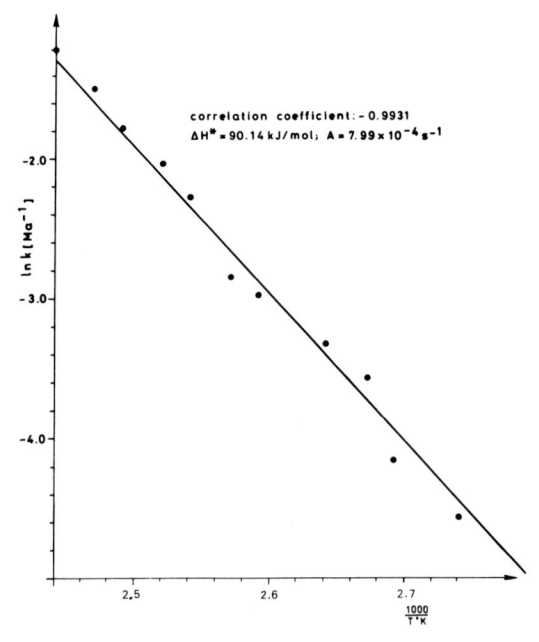

Fig. 10. Arrhenius plot of sterane isomerization in the case of calculation with the geological age of samples

Let us summarize the above findings: the enthalpy of activation from different calculation methods hardly varies, but the value of A gives a variation of some order of magnitude. The value of ΔS^* calculated from A has a considerable negative value in each case and this suggests that the activated complex is "tight".

The parameters were also computed by the most exact method of absolute time (Fig.11). The dispersion for A and ΔS^* was the least when we used the value of ΔH^* of 91.61 ± 1.47 kJ mol^{-1}. The results of this computation are given in Table 2. The value of A' of $1.899 \pm 0.226 \cdot 10^9$ Ma^{-1} ($6.022 \cdot 10^{-5}$ s^{-1}) was found according to method of absolute time. The value of ΔS^* of -278.3 J mol^{-1} °K^{-1} was obtained from A' (A = 0.0024 s^{-1} at 400° K).

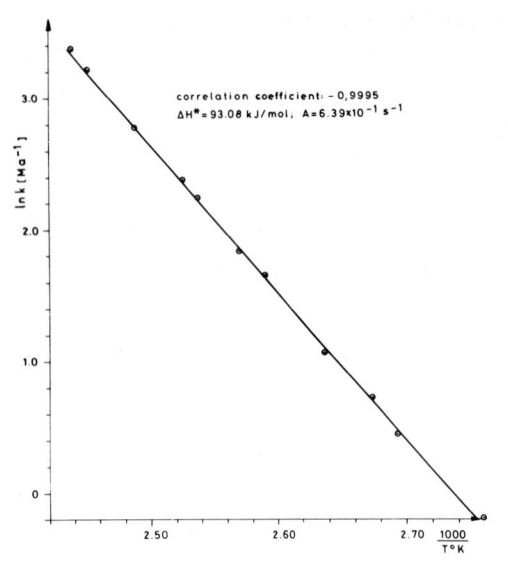

Fig. 11. Arrhenius plot of sterane isomerization calculated from the most exact method of absolute time (MAT) which yields the best fitting of data in this case

correlation coefficient: -0,9995
$\Delta H^* = 93.08$ kJ/mol, $A = 6.39 \times 10^{-1}$ s^{-1}

Table 2. Calculation of sterane isomerization with the method of absolute time (MAT)

$$A' = \frac{-1/\beta \ln(1-\beta\alpha)}{\int_0^\tau f(\tau) \cdot \exp(\frac{-\Delta H^*}{Rf(\tau)}) d\tau}$$

$f(\tau) = T = 238.10 + 16.34 \tau$; $\beta = 1.724$; $\Delta H^* = 91.61$ kJ/m

T_{p^*}			$-1/\beta \ln(1-\beta\alpha)$	$\int_0^\tau f(\tau) \exp \frac{-\Delta H^*}{Rf(\tau)} d\tau$	A'	$\ln k_{p^*}$	$-1/T_{p^*} \cdot 10^3$
4.90	362.3	0.05	0.0523	$2.6724 \cdot 10^{-11}$	$1.9570 \cdot 10^9$	-0.1975	2.760
5.38	371.3	0.08	0.0861	$5.3452 \cdot 10^{-11}$	$1.6108 \cdot 10^9$	0.4549	2.693
5.58	374.3	0.14	0.1602	$7.0774 \cdot 10^{-11}$	$2.2635 \cdot 10^9$	0.7188	2.672
5.85	379.3	0.18	0.2155	$1.0261 \cdot 10^{-10}$	$2.1001 \cdot 10^9$	1.0680	2.636
6.33	386.3	0.25	0.3271	$1.9472 \cdot 10^{-10}$	$1.6799 \cdot 10^9$	1.6694	2.589
6.47	389.3	0.28	0.3823	$2.3363 \cdot 10^{-10}$	$1.6363 \cdot 10^9$	1.8403	2.569
6.82	394.3	0.41	0.7117	$3.6515 \cdot 10^{-10}$	$1.9491 \cdot 10^9$	2.2591	2.536
6.93	396.3	0.46	0.9137	$4.1911 \cdot 10^{-10}$	$2.1801 \cdot 10^9$	2.3883	2.523
7.27	402.3	0.51	1.2262	$6.3694 \cdot 10^{-10}$	$1.9252 \cdot 10^9$	2.7803	2.486
7.67	408.3	0.55	1.7171	$1.0279 \cdot 10^{-9}$	$1.6706 \cdot 10^9$	3.2282	2.449
7.82	410.3	0.57	2.3526	$1.2253 \cdot 10^{-9}$	$1.9200 \cdot 10^9$	3.3925	2.437

$A' = (1.8993 \pm 0.2262) \cdot 10^9$; mêv $= 6.022 \cdot 10^{-5}$ s^{-1} → $\Delta S^* = -278.3$ J/m^o K; p^* = present

For comparison we give data from MACKENZIE and McKENZIE (1983), namely $E_a (\approx \Delta H^*) = 91$ kJ mol^{-1}, $A = 0.006$ s^{-1} and $\Delta S^* = -282$ J mol^{-1} °C^{-1}. They used the value of K (equilibrium constant) of 1.174 for calculations which were based on effective heating time estimations.

2) Isomerization of hopane at C-22

The equilibrium constant from the studied sample suite gives $K = 1.326$ and consequently, β is 1.754. The temperature-geological age relationship for our case can be characterized as $f(\tau) = T = 283.1 + 16.34\tau$ where T is the temperature of the sample in °K and τ is its age i.e. the time which has elapsed since burial in Ma.

The results of the effective heating time method are shown in Fig. 12 ($\tau_{EHT} = 1.16$ Ma). Considering that, H-I/7 sample has already reached equilibrium practically, calculations were also carried out with and without it. The first calculation yielded $\Delta H^* = 113.76$ kJ mol^{-1}, $A = 3.3$ s^{-1} and $\Delta S^* = -209.8$ J mol^{-1} °K^{-1} and the second gave $\Delta H^* = 96.09$ kJ mol^{-1}, $A = 3.3 \cdot 10^{-1}$ s^{-1} and $\Delta S^* = -256.0$ J mol^{-10} °K^{-1}.

The results of the differential method were also plotted in Fig. 12. This method yielded $\Delta H^* = 102.7$ kJ mol^{-1}, $A = 4.25$ s^{-1} and $\Delta S^* = -234.7$ J mol^{-1} °K^{-1}.

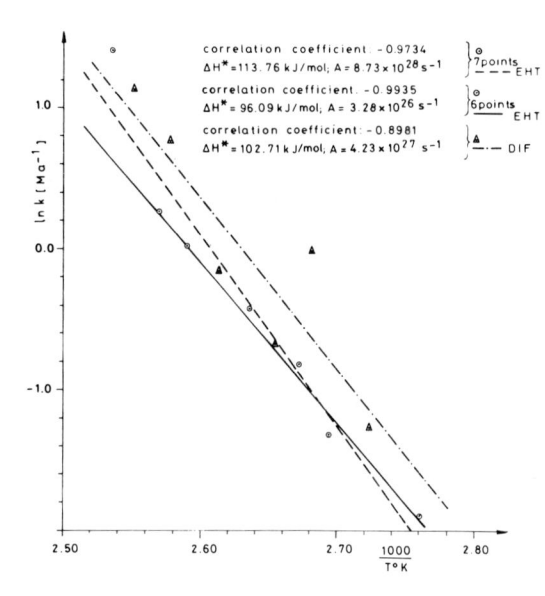

Fig. 12. Arrhenius plot of the hopane isomerization at C-22 calculated from methods of effective heating time (EHT) and differential (DIF). The dispersion of the second method is considerable again. The deepest sample (top position in the figure) has already reached equilibrium. The correlation coefficient was calculated in case of EHT method with and without equilibrium, too.

The reaction parameters were calculated using the geological age as duration, and the results are plotted in Fig. 13. The parameters obtained by this approach are: $\Delta H^* = 87.54$ kJ mol^{-1}, $A = 1.26 \cdot 10^{-3}$ s^{-1} and $\Delta S^* = -234.7$ J mol^{-1} °K.

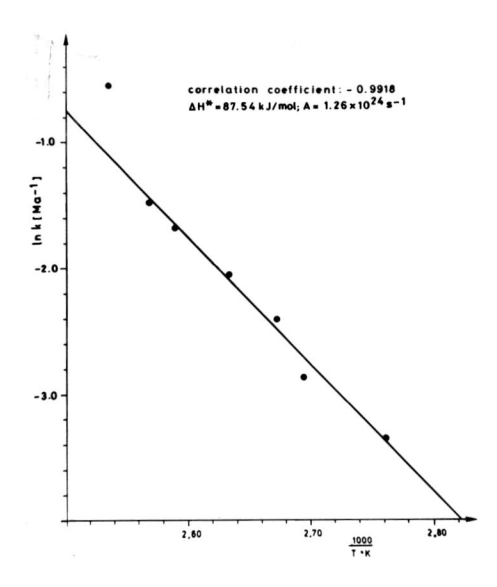

Fig. 13. Arrhenius plot of hopane isomerization calculated using the geological age of the samples as exposure time

The values for the activation energy obtained by different methods of approximation do not show too great a dispersion, nevertheless they scatter to a greater extent than those for sterane isomerization as a consequence of less survey data (Table 1), and therefore the dispersions of the pre-exponential factors and of the activation entropies are very considerable.

According to the method of absolute time calculations (Fig. 14 and Table 3) we obtained the least dispersion for the pre-exponential factor using the value $\Delta H^* = 87.54$ kJ mol^{-1}. The value of A' of $9.306 \cdot 10^{-5}$ s^{-1} $\pm$ $8.6 \cdot 10^{-7}$ s^{-1} and this means that $A = 3.5 \cdot 10^{-2}$ s^{-1} and $\Delta S^* = -274.7$ J mol^{-1} °K^{-1}.

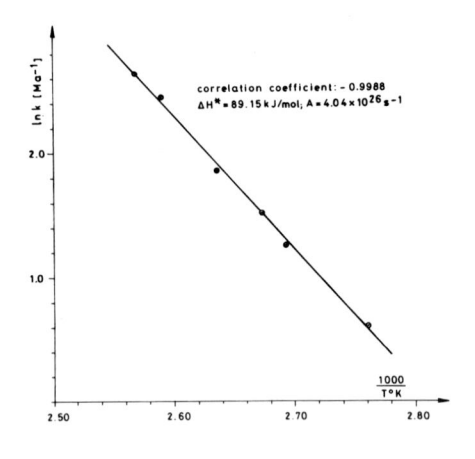

Fig. 14. Arrhenius plot of hopane isomerization calculated using method of absolute time (MAT), the least dispersion was received in this case (not included H-I/7 sample, which was nearly in equilibrium, see Table 1).

Table 3. Calculation of hopane isomerization with the method of absolute time (MAT)

$$A' = \frac{-1/\beta \ln(1-\alpha\beta)}{\int_0^\tau f(\tau) \cdot \exp(\frac{-\Delta H^*}{R\ \tau}) d\tau}$$

$f(\tau) = T = 283.10 + 16.34\ \tau$; $\beta = 1.754$; $\Delta H^* = 87.54$ kJ/mol

	T_{p*}	$-1/\beta \ln(1-\beta)$	$\int_0^\tau f(\tau)\exp\frac{-\Delta H^*}{Rf(\tau)}d\tau$	A'	$\ln k_{p*}$	$-1/T_{p*} \cdot 10^3$
4.90 0.15	362.3	0.1741	$6.1096 \cdot 10^{-11}$	$2.8496 \cdot 10^9$	0.6194	2.760
5.38 0.24	371.3	0.3155	$1.2134 \cdot 10^{-10}$	$2.6001 \cdot 10^9$	1.2643	2.693
5.58 0.34	374.3	0.5172	$1.6019 \cdot 10^{-10}$	$3.2286 \cdot 10^9$	1.5250	2.672
5.85 0.42	379.3	0.7608	$2.3136 \cdot 10^{-10}$	$3.2884 \cdot 10^9$	1.8698	2.636
6.33 0.50	386.3	1.1974	$4.3577 \cdot 10^{-10}$	$2.7416 \cdot 10^9$	2.4635	2.589
6.47 0.53	389.3	1.5130	$5.2176 \cdot 10^{-10}$	$2.8998 \cdot 10^9$	2.6322	2.569

$A' = (2.9347 \pm 0.2717) \cdot 10^9 = 9.306 \cdot 10^{-5}$; $\Delta S^* = -274.7$ J/mol °K; p^* = present

For comparison, MACKENZIE and McKENZIE (1983) obtained the following values: $E_a = 91.0$ kJ mol^{-1}, $A = 1.6 \cdot 10^{-2}$ s^{-1} and $\Delta S^* = -271$ J/mol °K. They used an equilibrium constant K of 1.564 for calculations which were based on the effective heating time of 6 Ma.

3) Aromatization of C-ring monoaromatic steroid hydrocarbons

The aromatization is an irreversible reaction and therefore the value of β is 1.

We applied different calculation models (Figs. 15, 16, 17 and 18) to aromatization. It is striking that the data in the Arrhenius plot are very far from being on a straight line. However, as was mentioned previously, the reasons for this phenomenon could be as follows:

i) The kinetics of the reactions cannot be described by a first-order equation.

ii) The cracking of aromatic steroid hydrocarbons were observed in sediments (SAJGÔ et al., 1983, 1984, 1985). ABBOTT et al. (1984, 1985) found that the concentration time function of triaromatic steroid hydrocarbons passes through a maximum in laboratory simulations. If the monoaromatic steroid hydrocarbons were more stable on the side-chain cleavage than triaromatic ones we would explain the anomalous fall-off of the aromatization rate constant at higher temperatures as is shown in Figures 15, 16 and 17, but there is no evidence of this.

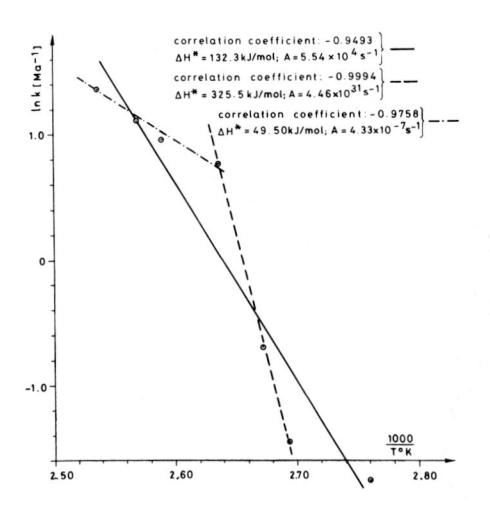

Fig. 15. Arrhenius plot of the conversion of C-ring monoaromatic to ABC-ring triaromatic steroid hydrocarbons calculated using the method of effective heating time. Calculations were performed in three different ways because of the great dispersions of the observation points.

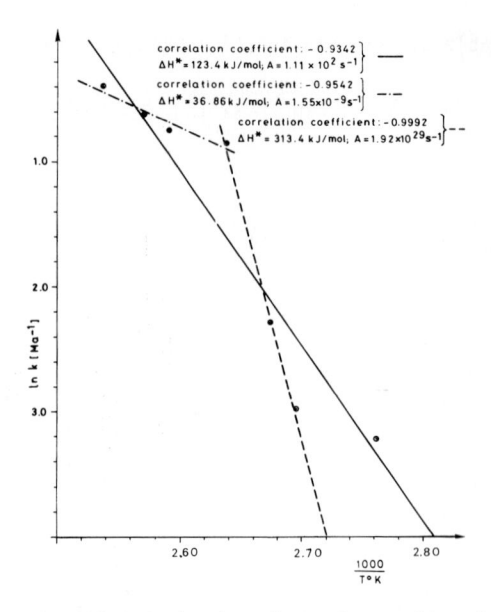

Fig. 16. Arrhenius plot of aromatization calculated using the whole geologic age of samples as duration. Calculations were performed for three cases because of the bad fit of the survey data.

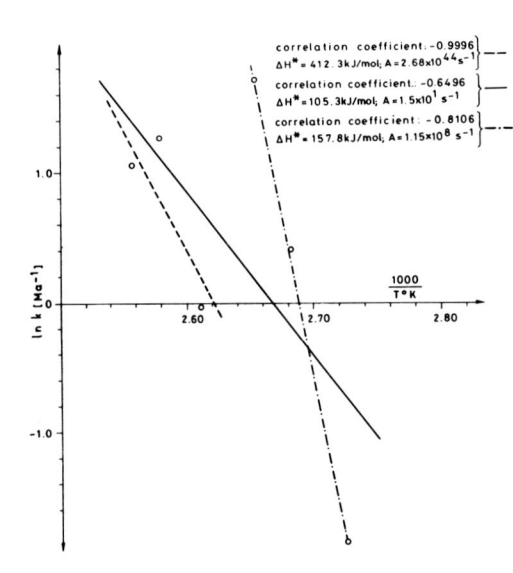

Fig. 17. Arrhenius plot of aromatization calculated using differential method. The dispersion is extremely great in this case, caused by insufficient observation points.

iii) The concentrations of reactants and products in samples are governed not only by the chemical reaction, but also by the rate of mass transfer, e.g. migration.

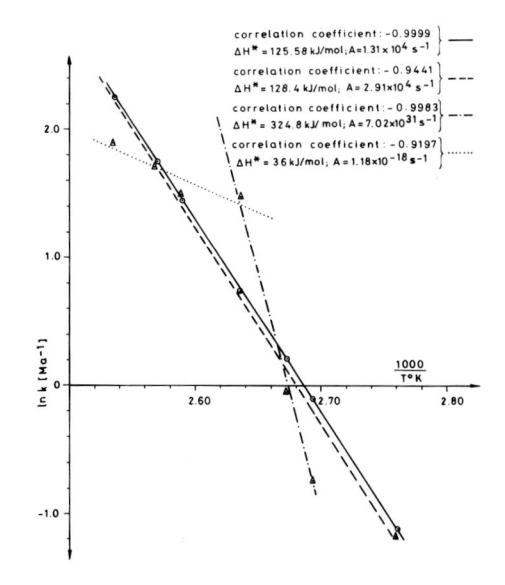

Fig. 18. Arrhenius plot of studied aromatization calculated using method of absolute time (MAT). Calculations were performed in three ways. Using ΔH* = 125.58 kJ mol^{-1} and A = 1.313·10^4 s^{-1} values, calculated points were represented as open circles in the plot.

iv) The deviations resulted from the inaccuracy of concentration measurements. The first survey datum falls into the range of low concentrations, and the last four survey data belong to the range of over 90 %. It is easy to see that small measuring errors in the range of the low and high conversions can produce considerable differences in the values of the rate constant.

Because of the great dispersion of survey data we computed the rate parameters for three different cases with all the methods applied. Firstly, we considered all the data, secondly, we used only the data in the medium, and thirdly we took the data from the range of high conversion into consideration.

The results are as follows:

a) i) all data with the effective heating time method:
$\Delta H^* = 132.3$ kJ mol^{-1}, $A = 5.54 \cdot 10^4$ s^{-1} and $\Delta S^* = -156$ J mol^{-1} °K^{-1};

ii) all data with the differential method:
$\Delta H^* = 105.3$ kJ mol^{-1}, $A = 15$ s^{-1} and $\Delta S^* = -224$ J mol^{-1} °K^{-1};

iii) all data with the geological age:
$\Delta H^* = 123.4$ kJ mol^{-1}, $A = 1.11 \cdot 10^2$ s^{-1} and $\Delta S^* = -207.6$ J mol^{-1} °K^{-1};

iv) all data with the absolute time method:
$\Delta H^* = 125.4$ kJ mol^{-1}, $A = 1.25 \cdot 10^4$ s^{-1} and $\Delta S^* = -168.3$ J mol^{-1} °K^{-1}.

b) i) The effective heating time method from medium conversions:
$\Delta H^* = 325.5$ kJ mol^{-1}, $A = 4.46 \cdot 10^{31}$ s^{-1} and $\Delta S^* = 359$ J mol^{-1} °K^{-1};

ii) the differential method from medium conversions:
$\Delta H^* = 412.3$ kJ mol^{-1}, $A = 2.68 \cdot 10^{44}$ s^{-1} and $\Delta S^* = 603.7$ J mol^{-1} °K^{-1};

iii) the geological age from medium conversions:
$\Delta H^* = 313.4$ kJ mol^{-1}, $A = 1.92 \cdot 10^{29}$ s^{-1} and $\Delta S^* = 313.8$ J mol^{-1} °K^{-1};

iv) the absolute time method from medium conversions:
$\Delta H^* = 324.8$ kJ mol^{-1}, $A = 7.02 \cdot 10^{31}$ s^{-1} and $\Delta S^* = 313.8$ J mol^{-1} °K^{-1}.

c) i) The effective heating time method from high conversions:
$\Delta H^* = 49.5$ kJ mol^{-1}, $A = 4.33 \cdot 10^{-7}$ s^{-1} and $\Delta S^* = -368.6$ J mol^{-1} °K^{-1};

ii) the differential method from high conversions:
$\Delta H^* = 157.8$ kJ mol^{-1}, $A = 1.15 \cdot 10^{-8}$ s^{-1} and $\Delta S^* = -92.4$ J mol^{-1} °K^{-1};

iii) the geological age from high conversions:
$\Delta H^* = 36.9$ kJ mol^{-1}, $A = 1.55 \cdot 10^{-9}$ s^{-1} and $\Delta S^* = 415.4$ J mol^{-1} °K^{-1};

iv) the absolute time method from high conversions:
$\Delta H^* = 36$ kJ mol^{-1}, $A = 1.18 \cdot 10^{-18}$ s^{-1} and $\Delta S^* = -590$ J mol^{-1} °K^{-1}.

The rate parameters can be divided in the next three groups according to conversion ranges:

i) for the total range:
$\Delta H^* = 121.6 \pm 11.5$ kJ mol^{-1}, $A = 1.06 \cdot 10^{3}$ s^{-1} and $\Delta S^* = -189 \pm 32$ J mol^{-1} °K^{-1};

ii) for the range of medium conversions:
$\Delta H^* = 344 \pm 46$ kJ mol^{-1}, $A = 2.0 \cdot 10^{34}$ s^{-1} and $\Delta S^* = 410 \pm 131$ J mol^{-1} °K^{-1};

iii) for the range of high conversions:
$\Delta H^* = 70 \pm 59$ kJ mol^{-1}, $A = 5.6 \cdot 10^{-7}$ s^{-1} and $\Delta S^* = -367 \pm 206$ J mol^{-1} °K^{-1}.

MACKENZIE & McKENZIE (1983) published the next values for aromatization:
$E_a = 200$ kJ mol^{-1}, $A = 1.8 \cdot 10^{14}$ s^{-1} and $\Delta S^* = 11$ J mol^{-1} °K^{-1}.

In the case of aromatization they also used the data of laboratory experiments in contrast with isomerizations of sterane and hopane. They obtained the rate parameters using extreme survey data. One end of the range comes from geological observations with a temperature range of 88.5 - 115° C and the other is derived from laboratory observations (250 - 260° C).

Summary

The concentrations of certain biological marker compounds were determined from core samples of the Hód-I borehole (Pannonian basin). The age and temperature data were available so we could compute the rate parameters from concentration data for the isomerizations of the sterane at C-20 ($\Delta H^* = 91.6 \pm 1.5$ kJ mol^{-1}, $A = 2.4 \cdot 10^{-3}$ s^{-1} and $\Delta S^* = -278.3$ J mol^{-1} °K^{-1}) and of the hopane ($\Delta H^* = 87.5$ kJ mol^{-1}, $A = 3.5 \cdot 10^{-2}$ s^{-1} and $\Delta S^* = -275$ J mol^{-1} °K^{-1}) and for aromatization of monoaromatic steroids (the average values: $\Delta H^* = 121.6 \pm 11.5$ kJ mol^{-1}, $A = 1.06 \cdot 10^3$ s^{-1} and $\Delta S^* = -189 \pm 32$ J mol^{-1} °K^{-1}). MACKENZIE & McKENZIE (1983) obtained the following parameters for the same reactions: $E_a (\approx \Delta H^*) = 91$ kJ mol^{-1}, $A = 6 \cdot 10^{-3}$ s^{-1} and $\Delta S^* = -282$ J mol^{-1} °K^{-1}; $E_a(\approx \Delta H^*) = 91$ kJ mol^{-1}, $A = 1.6 \cdot 10^{-2}$ s^{-1} and $\Delta S^* = -271$ J mol^{-1} °K^{-1}; $E_a(\approx \Delta H^*) = 200$ kJ mol^{-1}, $A = 1.8 \cdot 10^{14}$ s^{-1} and $\Delta S^* = 11$ J mol^{-1} °K^{-1}, respectively. Our parameters for isomerizations are in good agreement with those of MACKENZIE & McKENZIE (1983).

In the case of aromatization the deviations are remarkable between different authors and the results depend on the conversion ranges of the computations too. From this we have to conclude that in contrast to the isomerization, the aromatization studied is not known well enough yet to be applied in geothermal reconstruction.

On the basis of reaction studies, the time bounds and the confidence limit of the reconstruction of thermal and subsidence history are governed by the rate parameters and their authenticity. This problem is discussed in detail by LEFLER & SAJGÓ (1985).

Acknowledgements

We thank the National Oil and Gas Trust of Hungary, especially Dr. V. Dank, for providing samples and permission to publish. Cs. S. thanks Prof. L. Stegena for inviting him to present this paper at the Paleogeothermics Session of IASPEI General Assembly, which was held in Hamburg, FRG, August 15-27, 1983. Cs. S. also thanks Prof. W. Torge and the National Committee of Geodesy and Geophysics of the FRG, for covering the expenses of his participation in the Conference. The GC-MS runs were mainly carried out at OGU, School of Chemistry, Univ. of Bristol, U.K. during the tenure of Cs. S.'s fellowship of the Scientific Exchange Agreement of ESF. The Natural Environment Research Council supported the GC-MS facilities (GR3/3758). Cs. S. is grateful to Prof. G. Eglinton (Univ. of Bristol) and Prof. G. Guichon (Ecole Polytechnique Paris) for arranging the fellowship. Technical assistance from Mrs. A. Marôt and Mrs. V. Csontos in Budapest and Mrs. A. Gowar at Bristol is gratefully acknowledged. We are especially grateful to Prof. G. Eglinton and to Drs. G.D. Abbott, K. Raksány, G. Várhegyi and O. Tomschey for their critical review of the manuscript.

LIMITS OF APPLICATION OF THE REACTION KINETIC METHOD IN PALEOGEOTHERMICS

LEFLER, J. and C.S. SAJGÓ
Laboratory for Geochemical Research
Hungarian Academy of Sciences, Budaörsi út 45, H-1112 Budapest

Abstract

The applicability of organic geochemical reactions in thermal history reconstruction of basins is investigated. The requirements of kinetic parameters of the reactions suitable to study the basins of temporally increasing temperature, are discussed in detail. Two basin types with extreme heating rates, i.e. a rapid and a slow heating rate, are discussed. The thermal history of the basins is studied on the basis of a simplified model, in which the heating rate proved to be a constant value before τ_0 moment. At that moment, the heating rate was suddenly increased to the recent value.

The mathematical description of the model is given. Applying this to reactions of different ΔH^* activation energy and of different $\underline{A}$ pre-exponential factor it is demonstrated that to study the basins of increasing temperature the reactions of low activation energy and of low pre-exponential factors are more suitable to follow the thermal history. In the case of a slow heating rate, these reactions are suitable to trace the thermal history up to several 10 million years while in the case of a high heating rate this is possible only to several million years if the reaction parameters are exactly known.

The uncertainties deriving from the inaccuracies of measurement data are also studied, i.e. how these affect the time interval to be determined.

Finally, it is stated that in order to carry out a reliable thermal history reconstruction, the investigation should include the exact determination of parameters of the reactions of vitrinite-type, i.e. of low activation energy and of small pre-exponential factor. Reliable reconstruction free of contradictions can be carried out only in possession of a suitable number of data and by also taking into account the error calculations.

Introduction

In the previous work the reactions of three biological marker compounds were discussed and their kinetic parameters were determined (SAJGÓ & LEFLER, this volume). Nevertheless, the effect of heat flow changes in the sedimentary basins on the reactions and on the path of the chemical reaction in general was not dealt with.

In this paper we try to indentify those chemical reactions whose investigation makes possible the thermal history reconstruction of the evolution of a basin.

The rate of chemical reactions is described by the relationship

(1) $\quad r = \dfrac{dc}{d\tau} = A \cdot e^{-\Delta H^*/RT} \cdot f(c)$

i.e. in addition to the concentrations of the components partaking in the reaction, the reaction rate is determined by two constant parameters of the reaction rate: by the A pre-exponential factor and by the ΔH^* activation enthalpy. In the expression of the rate constant

(2) $\quad k = A \cdot e^{-\Delta H^*/RT}$

temperature is also involved, thus to follow the reactions this expression is also involved in the thermal history reconstruction.

In the following, it is assumed that the parameters defining the rate constant of the reactions to be discussed are universal in the sense that they are the same in all basins. In other words, the reaction mechanism is independent of the surrounding matrix. Under this condition two groups of questions can be answered:

1) In the case of basins of different types of thermal history, what are the A and ΔH^* parameters of the reactions which are most sensitive to the thermal changes?

2) The temporal progress of reactions can be followed by the analytical investigation of the reaction products. The parameters of the reaction are also known from measurement results. Consequently, all the available data bear some measurement errors. What is the level of difference from the theoretical values which is due to the changes in the thermal history and not due to the measurement errors?

The mathematical model

It is obvious from the expression (2) that the smaller the pre-exponential factor of the reaction studied, the slower the reaction under the same conditions, i.e. by studying the reaction products, the events of a larger time span can be followed. The value of ΔH^* refers to the thermal sensitivity of the reaction rate. If this value is low, the reaction rate is only slightly temperature dependent. If the value is great, the reaction rate equals practically zero at lower temperatures, i.e. the reaction is frozen. Consequently, the thermal history of the cooling basins can be followed by reactions of low pre-exponential factor and of high activation enthalpy. When in a basin, in a sequence of lower temperature, this type of reaction product can be identified in large amounts, one can be sure that the sequence in question endured a high temperature during geological times which enabled the reaction to proceed. The time span can be determined on the basis of reaction progress. (When drawing the conclusion, it should be taken carefully into account whether the products are autochthonous.)

The processes proceeding in basins of increasing temperature as a function of time can be easily followed. Let us take the very simple thermal history model below.

In the basin the temperature of the material under deposition and burial is linearly increasing, the rate being $\delta\,°C\,Ma^{-1}$. Assume that the sudden change followed before τ_0 million years (thermal catastrophe: e.g. increase of the geothermal gradient). As a consequence of this, on the one hand, the temperature of strata will suddenly increase, and on the other, since that time, the temperature of the subsiding material will increase faster. According to this simplified model let the rate of increase be $\gamma\,°C\,Ma^{-1}$, and also be linear. The thermal history of the subsiding material is shown in Fig. 1 according to this model. The temporal change of temperature of the reacting matter is:

(3) $\begin{cases} T = T_0 + \delta\tau & \text{if } \tau > \tau_0 \\ T = T_0' + \gamma\tau_0 & \text{if } 0 < \tau < \tau_0 \end{cases}$

Let the rate equation of the chemical reaction proceeding in the subsiding material be

(4) $\quad r = \dfrac{dB}{d\tau} = \vec{K}_A C_A - \overleftarrow{K}_B C_B$

(see: SAJGÓ & LEFLER, 1985). The solution of the reaction's differential equation is:

(5) $\quad -1/\beta \ln(1-\beta\alpha) = \displaystyle\int_0^\tau \vec{k}\,d\tau$

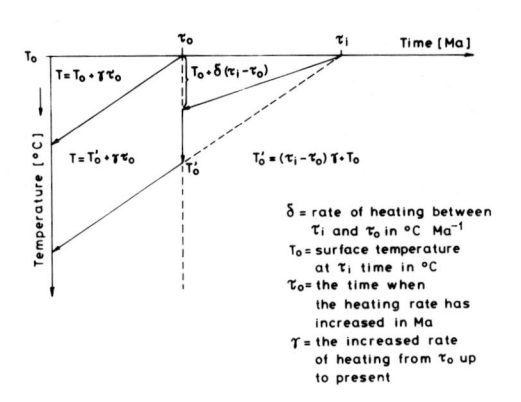

Fig. 1. A simplified thermal history model, serving as a basis for mathematical description. The heat flow, as the model describes, increases conforming to unit-jump function in basins at τ_0 moment.

The left hand side of the equation is the concentration dependent part. $\beta = 1$ if the reaction is irreversible and $\beta = \frac{K+1}{K}$ if the reaction can be characterized by an equilibrium constant, K and is a reversible reaction leading to equilibrium. α denotes the conversion (degree of transformation) that can be analytically determined. The right hand side of the equation depends on time, and through $\vec{k}$, also on the temperature. If one knows the $T(\tau)$ function, the integral can be solved. In the case of our simplified model this function is given by the relationship (3).

It is to be determined how the conversion values denoted by α_m, which can be calculated according to the temperature-time relationship of our thermal history model, differ from the α_c conversion values which can be calculated under the following conditions: in the basin the rate of heating of the subsiding samples would be continuously γ °C Ma^{-1}, i.e. in the basin the recent conditions would always predominate and the temperature gradient would be constant. If the difference between the α_m and α_c conversion values is higher than the double of the uncertainty deriving from the inaccurate measurement of the conversion values, then on the basis of this difference a thermal "catastrophe" (i.e. a rapid change) took place in the basin. If the difference is less than the double value, then no conclusion can be drawn of such a rapid change.

The thermal history model demonstrated in Fig. 1 seems to be rather rare in nature. In the case of thermal histories similar to that outlined in this figure, i.e. possessing only one sudden change, in nature the change cannot be described by a step-like discontinuous function, but rather by some continuous function. The heating curves preceding and following the thermal historical "catastrophe" are not necessarily linear. Nevertheless, the conclusions which can be drawn from this model are general since the change is sharpest in this model and if no change can be observed in the conversion caused by the reactions, then the conversion differences are not as great in the less sharp transitions.

In order to answer the first question of the introduction, let us take a chemical reaction of exactly known A pre-exponential factor and of exactly known ΔH^* activation energy, that can be described by equation (5).

If in the basin the thermal history has been undisturbed and could be characterized by the rate of heating of today (in Fig. 1, the section between the recent and τ_0), the relationship (5) is as follows:

$$(5a) \quad -1/\beta \ln(1-\beta\alpha_c) = 1/\gamma \int_{T_0}^{T_0+\gamma\tau} A \cdot e^{-\Delta H^*/RT} \, dT$$

If in τ_0 the change according to the model follows, the relationship (5) is modified:

$$(5b) \quad -1/\beta \ln(1-\beta\alpha_m) = 1/\delta \int_{T_0}^{T_0+\delta(\tau-\tau_0)} A \cdot e^{-\Delta H^*/RT} \, dt + 1/\gamma \int_{T_0'}^{T_0'+\gamma\tau_0} A \cdot e^{-\Delta H^*/RT} \, dT$$

In equation (5a) the calculated conversion, α_c (in the equation (5b) the measured conversion, α_m), is involved. Let us denote the difference of the calculated and measured conversions by m. Applying this denomination the conversion differences that can be obtained in two ways after τ time, can be solved by the expression below:

$$(6) \quad m = \alpha_c - \alpha_m = (\alpha_c - 1/\beta)(1 - e^{\Delta\beta})$$

where

$$(7) \quad \Delta = 1/\beta \ln \frac{1-\beta\alpha_m}{1-\beta\alpha_c} = A[1/\gamma \int_{T_0}^{T_0+\gamma(\tau-\tau_0)} \exp(\frac{-\Delta H^*}{RT})dT - \int_{T_0}^{T_0+\delta(\tau-\tau_0)} \exp(\frac{-\Delta H^*}{RT})dT]$$

Thus, our task is to study the changes of the mathematical relationship (6) according to A, ΔH^*, β and τ, and to search for the extreme values. The analysis of the relationship (6) is rather complicated. Nevertheless, some conclusions can be drawn concerning the formation of the function m as a function of reaction parameters and of the time.

1. The most conspicuous one is that the increase of the shock measure produces the progressive increase of difference (m) between the calculated and measured conversion.

2. Under the same conditions, the reactions of lower pre-exponential factor are more sensitive to the thermal jumps.

3. In basins of monotonously increasing temperature the reactions of lower activation enthalpy are more suitable to evaluate the thermal jumps.

4. With increasing β the sensitivity of the reaction reflecting the thermal jump will decrease. The irreversible reactions are more suitable to the analysis, than the reversible, equilibrium reactions.

5. It is obvious that the value m runs over a maximum as a function of τ. In the samples older than the change itself the deviation of the measured conversions from the calculated ones will increase as a function of the sample age, then after passing through a maximum value, it will decrease.

Natural boundary conditions

Equations (6) and (7) can be analyzed in the case of any γ and δ value-pair of the relationship (3). From the point of view of the numerical analysis the A, ΔH^* and β parameters of the chemical reactions can also be optional. Nevertheless, it is unequivocal that the rate of heating of the real geological basins endures some limitations and the known chemical reactions bear also definite A, ΔH^* and β triple regions.

The short comprehension of literature below will outline these conditions.

The fact that in this paper the application of the transformation of organic compounds in the thermal history, the temperature range to be studied is restricted. Most of the authors, e.g. LOPATIN (1971), BOSTICK (1973), HOOD et al. (1985) roughly agree that in nature the interpretable chemical reactions are completed up to 280 to 300° C, i.e. up to the formation of the anthracite state. If in our investigations only the transformations of biomarkers or the oil-generating reactions of kerogens are taken into account, the thermal maximum should be defined at lower temperature values. It is to be noted here that when tracing a rather wide temperature interval by some of the chemical reactions, special care should be attributed to this investigation. It may occur that the reaction path will change as a function of temperature, thus the equations describing the transformation itself will also change.

In a basin, the rate of heating depends on two factors: on the rate of subsidence of the basin, and on the geothermal gradient (heat flow) of the basin:

(8) $\quad dT/dt = (dz/dt) \cdot (dT/dz)$

where dz/dt, i.e. the rate of the subsidence of the basin is hypothesized to be between 20 and 400 m/Ma according to FISCHER (1969), KATZ (1979), SCHWAB (1976), TISSOT et al. (1980). In sedimentary basins the measure of the dT/dz geothermal gradient may vary between 20 to 45° C/km as a function of the thermal conductivity of rocks and of the heat flow (GRETENER, 1981).

Consequently, according to the relationship (10) the rate of heating may vary practically between 0.4° C Ma^{-1} and 18° C Ma^{-1}, i.e. the maximal value of ν is 18° C Ma^{-1} and the minimal value of δ is 0.4° C Ma^{-1}.

The organic geochemical literature refers to numerous reactions differing in their parameters. These are fairly exactly comprehended by the papers of SNOWDON (1979) and WAPLES (1984). Table 1 is derived from them. Unfortunately, the pre-exponential factors of the reaction are not listed, their measure can be calculated only from the column "Time Interval" when assuming that the reaction is completed if the conversion reaches 90 to 99 % of the maximal value. The estimated interval of the A value obtained in this manner is demonstrated in the last column of our table.

It is obvious from the table that both the activation enthalpies and the pre-exponential factors vary within a rather wide range. Out of the reactions published, the reaction natural coalification I bears the least pre-exponential factor ($A = 3.6 \cdot 10^{-12}$ Ma^{-1}) and the reaction pyrolisis IV is of the greatest value ($A = 1.8 \cdot 10^{47}$ Ma^{-1}). The difference is 59 orders of magnitude! The activation enthalpy of the reaction of natural coalification II of least activation enthalpy ($\Delta H^* = 3.8$ kJmol^{-1}) is only about 1/130 of the activation enthalpy of the reaction of kerogen pyrolisis I of highest activation enthalpy ($\Delta H^* = 485$ kJmol^{-1}).

To demonstrate the applicability of this wide scale the behavior of some reactions of different parameters will be demonstrated in two basins of extremely differing rate of heating.

Investigation of the applicability of reactions with different parameters according to the instances of basins with different rates of heating

Here we shall demonstrate how the reactions of different ΔH^* activation enthalpy and of different A pre-exponential factor react to a thermal "catastrophe" in basins having an extremely different rate of heating. According to the simplified model, the catastrophe denotes the increase of heat flow according to the unit-jump function, in all cases.

Table 1. Kinetic parameters for evolution of organic geochemical reactions

Reaction(s) observed	Temperature range °C	Time interval	Activation Energy kJmol^{-1}°K^{-1}	A s^{-1}	Reference
Rate of generation of bitumen	20-80	10-150 Ma	63	$10^8 - 2 \cdot 10^{13}$	TISSOT, 1969
Rate of generation of resins	20-80	10-150 Ma	59	$2.7 \cdot 10^7 - 4.0 \cdot 10^8$	
Rate of generation of hydrocarbons	20-80	10-150 Ma	84	$2.6 \cdot 10^8 - 2.4 \cdot 10^{12}$	
Kerogen pyrolysis I.	150-410	-	63-485	-	PETERS et al., 1977
Kerogen pyrolysis II.	350	-	189	-	
Oil-shale pyrolysis I.	350-437	-	62	$2.8 \cdot 10^{18}$	HUBBARD & ROBINSON, 1950
Oil-shale pyrolysis II.	437-525	-	25.4	$1.4 \cdot 10^7$	
Coal pyrolysis I.	350-550	-	25	$1.7 \cdot 10^{-3}$	van KREVELEN et al., 1951
Coal pyrolysis II.	350-550	-	82-122	67 - 1500	
Kerogen pyrolysis III.	150-200	5.1-101 hours	63	$1.06 \cdot 10^6 - 2.12 \cdot 12.10^{17}$	PETERS et al., 1977
Kerogen pyrolysis IV.	350-410	5-100 hours	481	$9.4 \cdot 10^{45} - 1.8 \cdot 10^{47}$	
Natural coalification I.	60-80	-	35	$3.6 \cdot 10^{-12}$	HUCK & KARWEIL, 1955
Natural coalification II.	140-160	-	3.8	0.3	KARWEIL, 1975
Natural oil generation	60-127	-	45-59	-	CONNAN, 1974
Oil cracking	267-372	-	206	$3 \cdot 10^{11}$	McNAB et al., 1952
Generation of methane from Green R. shale	185-400	$10^3 - 10^2$ a	169	$2 \cdot 10^6 - 4 \cdot 10^8$	HOERING & ABELSON, 1963
Changing proportion of gasoline naphthenes	32-145	9.7-480 Ma	23-26	$12 - 7.4 \cdot 10^2$	YOUNG et al., 1977
Changing proportion of 3 and 4-ring naphthenes	33-145	9.7-500 Ma	11-18	0.60-30	

One type of basin is the Pannonian-type basin in which recently the heating rate (γ) is roughly 16° C Ma^{-1}, thus δ proved to be 8° C Ma^{-1} before the date of the catastrophe (τ_0).

The other type (for simplicity let us term it a basin with an extremely low rate of heating) bears recently a heating rate of 1.6° C Ma^{-1}. Thus, before the τ_0, the rate of heating was only 0.8° C Ma^{-1} in this basin. With these two basins the ranges produced by nature are covered.

Calculations were carried out according to the relationship (6). Results, i.e. the differences between the true conversion and the conversion calculated after the recent rates of heating are plotted as a function of the age of samples.

First, the reaction of the sterane isomerization type reactions ($\Delta H^* = 90$ kJmol^{-1}; $A = 4.8 \cdot 10^8 - 10^{13}$ Ma) are demonstrated in Fig. 2 for thermal historical catastrophes of different τ_0 dates, in the Pannonian-type basin. Both in this figure as well as in the subsequent ones the limit value of m = 0.04 is shown which indicates the limit of reliability of the conversion measurements.

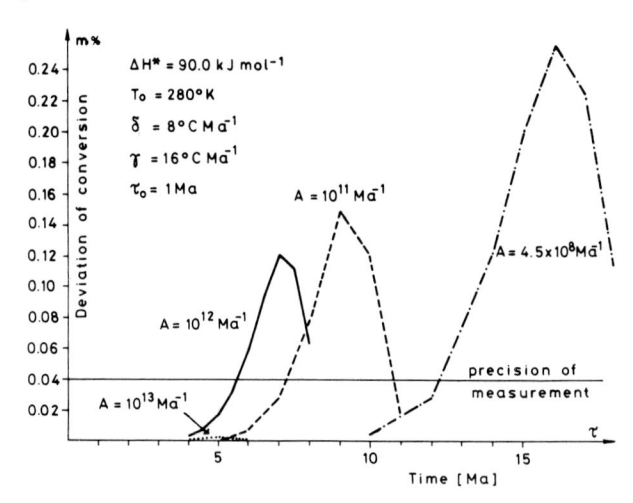

Fig. 2. Deviation of the calculated conversions from the measured ones in the case of sterane isomerization in a Pannonian-type basin if the heat flow doubled one million years ago.

Fig. 3 shows how the aromatization type reactions of highest activation enthalpy ($\Delta H^* = 150$ kJmol^{-1}) react to the sudden doubling of the rate of heating in the Pannonian-type basin.

Fig. 4 shows the behavior of a reaction of rather low activation enthalpy ($\Delta H^* = 10$ kJmol^{-1}) in a Pannonian-type basin. Such a reaction, lying on the boundary

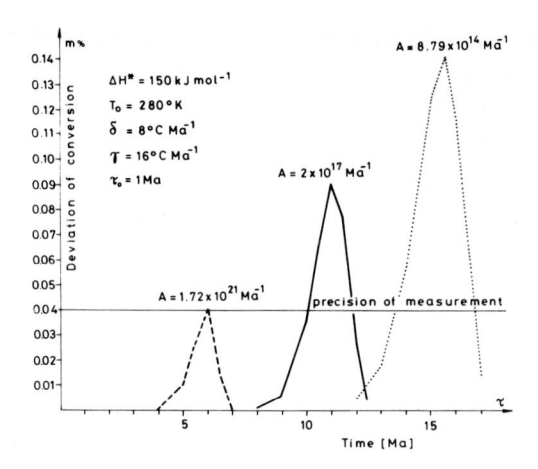

Fig. 3. Deviation of the calculated conversions from the measured ones in the case of aromatization-type reactions in a Pannonian-type basin if the heat flow doubled one million years ago.

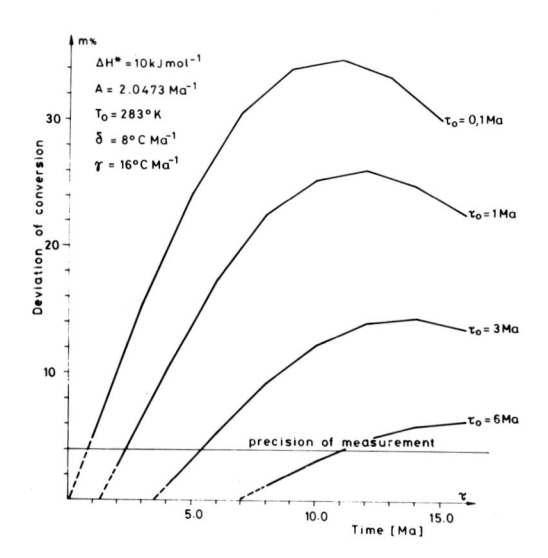

Fig. 4. Deviation of the calculated conversions from the measured ones in the case of vitrinite-type reactions in a Pannonian-type basin when heating rate doubled at some time.

of chemical reactions, is the reaction of "natural coalification I" in Table 1, its activation energy is only 3.8 kJmol^{-1}. In the following the reactions of such low activation energy will be termed vitrinite-type reactions. As it is shown in the figure, by means of this reaction rather distant, i.e. more than 6 million years, thermal jumps can be concluded if it was accompanied by the doubling of the heat flow. In

Figs. 5, 6 and 7 the applicability of the three reactions (isomerization, aromatization, vitrinite) are demonstrated in basins with a low rate of heating. Obviously, the investigation of a reaction of a given data-pair may give one an insight into the distant past in such a basin of low rate of heating, rather than in a Pannonian-type basin. When studying the heat flow doubling by means of the isomerization type reactions, conclusions can be drawn up to 10 million years (Fig. 5), by means of the aromatization type reactions (Fig. 6), up to 5-13 million years depending on the pre-exponential factor, and, by means of the vitrinite type reactions (Fig. 7), this time span can be enlarged up to 15 million years.

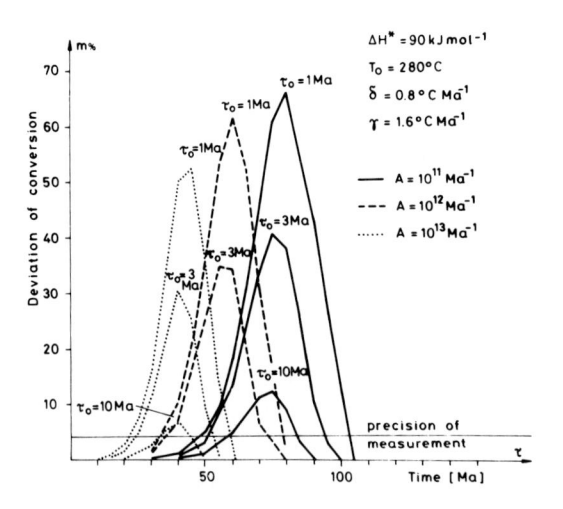

Fig. 5. Deviation of calculated conversions from the measured ones in the case of sterane isomerization-type reactions in basins with an extremely low heating rate when the heating rate doubled at some time (τ_0).

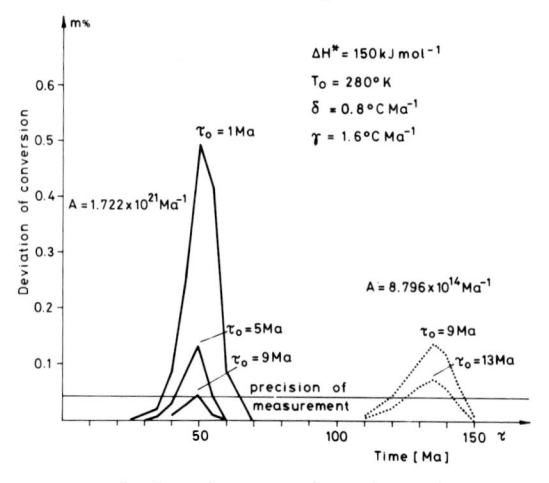

Fig. 6. Deviation of the calculated conversions from the measured ones in the case of aromatization-type reactions in basins with an extremely low heating rate when the heating rate was doubled at some time (τ_0).

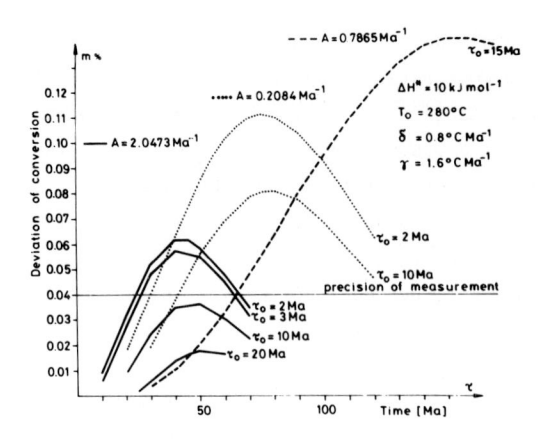

Fig. 7. Deviation of the calculated conversions from the measured ones in the case of vitrinite-type reactions in basins with an extremely low heating rate when the heating rate doubled at some time (τ_0).

Based on the figures, the overall conclusion can be drawn that out of the reaction of the same activation enthalpy those of lower pre-exponential factor are more favourable not only because these allow us to go back in more distant times, but because the lower the pre-exponential factor, the wider is the sample value interval from which reliable conclusions can be drawn to a thermal "catastrophe" (see e.g. Figs. 5 and 6). It can also be seen (e.g. Fig. 6) that in the case of reactions of lower pre-exponential factor the investigation of the older sample series, and, in the case of reactions of greater pre-exponential factor that of the relatively younger sample series, may throw light upon a possible change in the thermal history. Nevertheless, the pre-exponential factor of a reaction can be theoretically neither. If the pre-exponential factor of a reaction is very low, the range of the measurable conversions needs such long times which in a basin of monotonously increasing temperature as discussed above means such a high temperature that it falls out of the range of the organic chemical reactions. In our examples, the maximal temperature was defined between 200 and 250° C.

In Fig. 8 the behaviour of reactions as a result of the "catastrophe" are demonstrated, the reactions bearing the same feature that in the Pannonian-type basin produced conversion of 99 % during 14 million years, and the "catastrophe" in the thermal history characterized by the sudden doubling of heating rate τ_0 million years ago. The parameters of reactions are shown in Fig. 8, covering practically the whole of the range indicated in Table 1.

Similary to Fig. 8, in Fig. 9 the differences between conversions of undisturbed reactions and conversions of "catastrophe" disturbed reactions are demonstrated in a

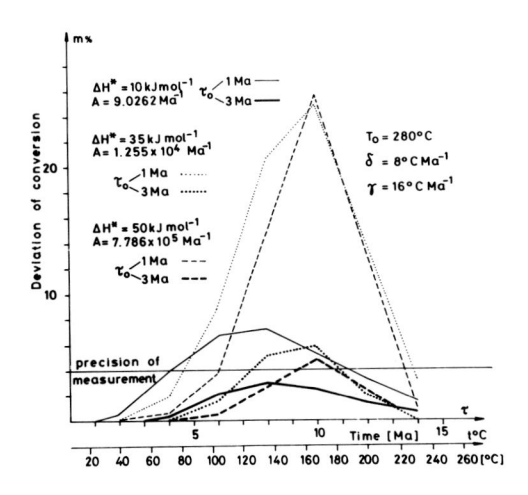

Fig. 8. Deviation of the calculated conversions from the measured ones in the examples of three different reactions in a Pannonian-type basin if the heating rate doubled one or three million years ago.

basin with an extreme low heating rate. The "catastrophe" was 10 million years ago, produced by the sudden doubling of the rate of heating. All these reactions have got conversion of 99 % during 100 million years, and their parameters are shown in Fig. 9.

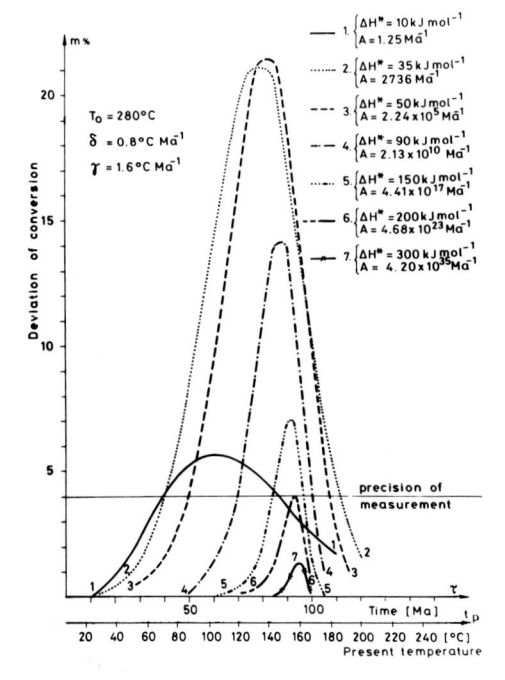

Fig. 9. Deviation of the calculated conversions from the measured ones in the examples of the reactions in basins with an extremely low heating rate if the heating rate doubled ten million years ago.

Finally, let us study how sensitive the reactions are when the rate of heating is increased only by a small extent. Let us have Table 2 as an example showing the progress of a vitrinite-type reaction of $\Delta H^* = 10$ kJmol^{-1} and $A = 2.04$ Ma^{-1} as a function of time when the rate of heating was doubled before 1 million years, i.e. from 8 to 10° C Ma^{-1}, and when it was increased only by 33.3 %, i.e. from 12 to 16° C Ma^{-1}.

Table 2. The variation of the deviation of conversion with different temperature jump.

τ(Ma)	2.00	4.00	6.00	8.00	10.00	12.00	14.00	16.00
α_c	0.0710	0.1664	0.2820	0.4092	0.5376	0.6568	0.7590	0.8402
m33	0.0701	0.1584	0.2606	0.3711	0.4835	0.5913	0.6889	0.7726
m33 %	0.09	0.80	2.14	3.81	5.41	6.55	7.01	6.76
m100	0.0692	0.1507	0.2400	0.3338	0.4285	0.5208	0.6074	0.6861
m100 %	0.17	1.57	4.20	7.54	10.91	13.60	15.16	15.41

τ = time (Ma); α_c: calculated conversion; m33: measured conversion in the case of temperature enhancement of 33 %; m33 %: deviation of the calculated conversion from the measured values; m100: measured conversion in the case of temperature enhancement of 100 %; m100 %: deviation of the calculated conversion from the measured values.

Having analyzed the data from Table 2, the difference between the measured and calculated conversion values (m) increases roughly parallel with the increase in the rate of heating measured in °C/Ma.

Summing up: based on the listed examples and in harmony with the conclusions derived from the mathematical model it can be stated that:

1) The reactions of lower activation energy and of lower pre-exponential factor are more suitable to study the thermal history of monotonously increasing temperature than the reactions of higher parameters.

2) The low pre-exponential factor is advantageous since parallel with the decreases of the pre-exponential factor, the range of measurement point of particularly different conversions will increase. At the same time, the decrease of the pre-exponential factor is accompanied by the increase of sample ages.

3) The differences between the measured and calculated conversion values (m) increase roughly proportionally at around the maximum with the change of the rate of heating at the moment of the "shock", measured in °C/Ma^{-1}.

4) In basins of extreme high heating rate (Pannonian-type basin) the thermal history can be traced back to about 6 million years when studying the organic chemical

reactions. This value can be approximated by studying the so-called vitrinite-type reactions ($\Delta H^* = 10$ kJmol^{-1}). This value is about 1 to 2 million years in the case of the isomerization-type reactions ($\Delta H^* = 90$ kJmol^{-1}), and taking into account the recent level of measurement techniques, to several hundred thousand years in the case of the aromatization-type reactions ($\Delta H^* = 150$ kJmol^{-1}).

5) In basins of extreme low heating rate the temporal limits of the reconstruction of thermal history are enlarged when studying the organic chemical reactions. In the case of reactions of low activation energy, the limiting values may be as great as several 10 million years. This wide temporal interval can also be studied by the reactions of high activation enthalpy. When applying the aromatization-type reactions the reconstruction of the thermal history of the last 10 million years can also be performed. In this type of basin the measure of the activation enthalpy has less effect on the investigation itself than in the basins with a rapid heating rate.

The basins of intermediary heating rate lie between these two extreme values from the point of view of the reconstruction of thermal history.

Limits of reconstruction of thermal history derived from the determination of reaction parameters

It has been demonstrated above how the reconstruction of the thermal history can be carried out in different basin types, i.e. the time intervals can be determined. These calculations are based on the fact that the reaction parameters (activation energy and pre-exponential factor) are considered to be known exactly, consequently the only limiting factor proved to be the accuracy of conversion measurements.

Unfortunately, this is not true in practice. It can also be seen in Table 1 that in many cases, not discrete values but rather intervals are given to the reaction parameters.

The effect of the error of pre-exponential factor and activation energy on the accuracy of the reconstruction of thermal history will be interpreted below, but in a somewhat different way than is usual.

The fact that in a basin the heating rate was changed during the geological times, can be concluded in our case when a reaction of known activation energy proceeded in the basin is studied. Assuming undisturbed thermal history and applying the relationship (5a), a series of pre-exponential factors can be calculated from the measured conversion from the age of the studied samples and from the activation enthalpy. In Table 3 several series are demonstrated. In our example the heat flow was doubled at τ_0 (the heating rates being δ and γ after and before the catastrophe, respectively). In the basins of Pannonian-type and of extreme low heating rate examples were given for two reaction types, i.e., for the vitrinite-type and for the sterane isomerization type reactions. It is shown by these examples that the pre-exponential factor sensitively follows the increase in the heating rate. The apparent pre-exponential factor calculated from the measured α_m data shows good agreement with the true A values up to the τ_0 point, then it decreases. Thus, the reaction seems to be slower than the true value. In many cases, however, the apparent pre-exponential factor is not less than the real pre-exponential factor!

Now, let us investigate the error in the calculation of the pre-exponential factor caused by the conversion measurements, temperature determinations, the determination of the geological times and by the error of the activation enthalpy, even in the case where the calculation performed takes a true thermal history into account.

The task is a simple calculation of error. Mathematically, the errors are:

the error from the determination of the geological age:

$$(9) \quad \left(\frac{\Delta A}{A}\right)_\tau \approx \frac{-T \exp\left(\frac{-\Delta H^*}{RT}\right)}{I} \cdot \Delta\tau$$

here and below:

$$(10) \quad I = \int_0^\tau T \exp\left(\frac{-\Delta H^*}{RT}\right) d\tau$$

the error from the inaccuracy of ΔH^*:

$$(11) \quad \left(\frac{\Delta A}{A}\right)_{\Delta H^*} \approx \frac{\int_0^\tau \exp\left(\frac{-\Delta H^*}{RT}\right) d\tau}{R \cdot I} \cdot \Delta(\Delta H^*)$$

the error from the inaccuracy of conversion measurements:

$$(12) \quad \left(\frac{\Delta A}{A}\right)_\alpha \approx \frac{\beta}{(1-\alpha\beta)\ln(1-\alpha\beta)} \cdot \Delta\alpha$$

the error derived from the inaccuracy of the determination of the equilibrium conversion:

$$(13) \quad \left(\frac{\Delta A}{A}\right)_\beta \approx 1/\beta \left[\frac{\alpha\beta - (1-\alpha\beta)\ln(1-\alpha\beta)}{\beta(1-\alpha\beta)\ln(1-\alpha\beta)}\right] \cdot \Delta\beta$$

Table 3. The variation of the pseudo-pre-exponential factor by the increase of heating rate

Basin with extreme high heating rate (Pannonian type): $\nu = 16°$ C/Ma; $\delta = 8°$ C/Ma; $T_o = 280$ K

1.) $\Delta H^* = 10.0$ kJmol^{-1}; $A_{real} = 2.047$ Ma^{-1};

τ (Ma)	1	2	3	4	6	8	10	12	14	16
α_m (%)	3.23	6.92	11.58	16.49	26.84	37.47	47.86	57.58	66.35	73.96
A_c (Ma^{-1})	2.05	2.07	2.04	2.06	1.995	1.880	1.774	1.681	1.601	1.532

2.) $\Delta H^* = 90$ kJmol^{-1}; $A_{real} = 10^{11}$ Ma^{-1};

τ (Ma)	4.00	5.00	6.00	7.00	8.00	9.00	10.00	11.00
α_m (%)	0.04	0.13	0.44	1.35	3.79	9.73	22.42	44.68
A_c (Ma^{-1})	2.88·10^{10}	2.11·10^{10}	1.82·10^{10}	1.6·10^{10}	1.43·10^{10}	1.29·10^{10}	1.17·10^{10}	1.08·10^{10}

3.) $\Delta H^* = 90$ kJmol^{-1}; $A_{real} = 10^{11}$ Ma^{-1}; $\tau_o = 1$ Ma

τ (Ma)	2.00	3.00	4.00	6.00	8.00	9.00	10.00	11.00
α_m (%)	0.01	0.03	0.15	2.29	20.63	47.31	80.80	98.12
A_c (Ma^{-1})	2.22·10^{11}	1.10·10^{11}	1.08·10^{11}	9.61·10^{10}	8.54·10^{10}	8.07·10^{10}	7.64·10^{10}	7.24·10^{10}

Basin with extreme low heating rate: $\nu = 1.6°$ C/Ma; $\delta = 0.8°$ C/Ma; $T_o = 280$ K

1.) $\Delta H^* = 10.0$ kJmol^{-1}; $A_{real} = 2.047$ Ma^{-1}; $\tau_o = 3$ Ma

τ (Ma)	5	10	20	30	40	50	60	70
α_m (%)	13.72	26.36	47.90	64.60	76.94	85.64	91.45	95.15
A_c (Ma^{-1})	2.038	1.990	1.888	1.795	1.711	1.636	1.569	1.511

2.) $\Delta H^* = 90.0$ kJmol^{-1}; $A_{real} = 2.13 \cdot 10^{10}$ Ma^{-1}; $\tau_o = 10$ Ma

τ (Ma)	40	60	70	75	80	85	90	95	100
α_m (%)	0.25	3.80	12.23	20.60	32.91	49.13	67.44	83.92	94.63
A_c (Ma^{-1})	1.800·10^{10}	1.607·10^{10}	1.540·10^{10}	1.507·10^{10}	1.475·10^{10}	1.444·10^{10}	1.414·10^{10}	1.384·10^{10}	1.354·10^{10}

the error from the inaccuracy of temperature measurements:

$$(14) \quad \left(\frac{\Delta A}{A}\right)_{T_{rec}} \approx - \frac{T \cdot \exp(\frac{-\Delta H^*}{RT})}{\gamma^2 \cdot I} \cdot \Delta T$$

In harmony with the error propagation laws the total relative error of the calculation of A:

$$(15) \quad \left(\frac{\Delta A}{A}\right) = \sqrt{\left(\frac{\Delta A}{A}\right)_\tau^2 + \left(\frac{\Delta A}{A}\right)_{\Delta H^*}^2 + \left(\frac{\Delta A}{A}\right)_\alpha^2 + \left(\frac{\Delta A}{A}\right)_\beta^2 + \left(\frac{\Delta A}{A}\right)_{T_{rec}}^2}$$

It follows from the numerical analysis of the individual components that the relative error of the error member from the geological age determination remains below 25 % if the relative error or age determination does not exceed 4 percent. The value of the error member derived from the inaccuracy of the determination of activation enthalpy varies around 30 percent if the uncertainty of the activation enthalpy is 1 percent, but is increased up to 170 % if the error of activation enthalpy increases to 5 %. Taking the recent determination inaccuracies of temperature determinations in boreholes, this error can be neglected.

The uncertainties in the α and β measurements may cause the greatest errors. According to MACKENZIE & McKENZIE (1983) the absolute error of the α determination may amount to 4 %. In Fig. 10 it can be shown how large errors are caused in the calculation of A in the case of the hopane and sterane isomerization reactions as well as of the aromatization reactions in different conversion ranges. It is seen in the figure that the error derived from the conversion measurements does not exceed the value derived from the other components if one takes the conversion range of 0.2 to 0.5 in the case of the isomerization reactions, and the conversion range of 0.15 to 0.95 in case of the aromatization reactions. (Taking into account that the two first reactions are reversible with an equilibrium conversion of 50 to 60 %, and that the third reaction is irreversible, this means practically the same ranges in the possible conversion degrees.)

Consequently, at the recent level of measurement technique the determination of the pre-exponential factor from each point may have an error of 45 to 175 percentage.

Comparing this fact with the data in Table 3 and with the sensitivity of the pre-exponential factor to the "catastrophes", it can be stated that the organic chemical reactions are suitable to the thermal history reconstruction if one possesses sufficiently large measurement numbers from the range of medium conversion! One must keep in mind that in this relation the error of the measurement series decreases proportionally to the number of members of the measurement series.

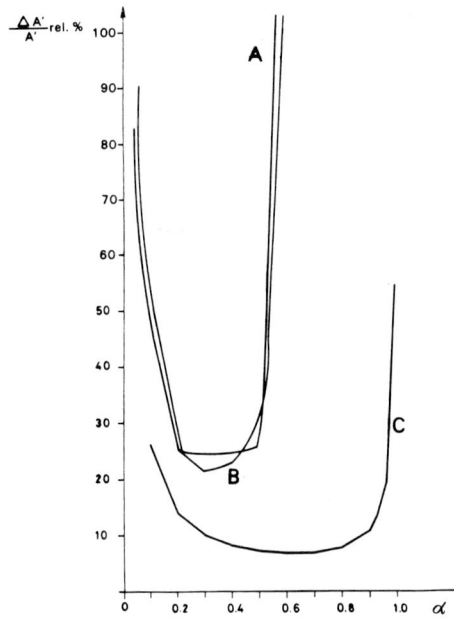

Fig. 10. The error of the calculated pre-exponential factors as a function of conversion if the precision of the conversion measurements is 4 percentage absolute. (A: sterane conversion, B: hopane conversion, C: aromatization).

When creating the thermal history reconstruction the error test should always be carried out. This is especially necessary when the determination of the reaction parameters is taken from natural data, since in these cases the errors may occur twice. If this is not carried out, one may have the fate, cited by LERCHIE, YARZAB & KENDAHL (1984) in connection with the application of vitrinite reflectance data, where the same data series may produce the reconstruction of many different thermal histories!

Summary, conclusion and suggestions

By means of the mathematical analysis of a simplified thermal history model it was demonstrated in which time intervals the reactions of different parameters (activation enthalpy and pre-exponential factor) are suitable to the thermal history reconstruction of a basin. In this paper the conditions were studied in detail under which, during the thermal evolution of a basin, the heating rate of the buried organic matter increased at one moment (this is the date of the "catastrophe") due to the increase either of the heat flow and/or of the rate of subsidence. It was demonstrated that a reaction is more suitable to study the distant moments, the lower is its activation energy. Out of the reactions of low activation energy those of lower pre-exponential factor are suitable to study the more distant time intervals. These are the reactions where the role of the time factor is greater than that of the temperature factor. On the contrary, as has been emphasized, the reactions sensitive to temperature (i.e. being of high activation energy) are more suitable to the thermal history reconstruction of cooling basins.

The applicability of reactions was studied in basins with two extremely differing heating rates. One of these is the Pannonian-type basin, the recent heating rate of which being 16° C Ma^{-1}, the other is the basin of extremely low heating rate, the recent heating rate being 1.6° C Ma^{-1}. It was demonstrated that in the Pannonian-type basins, the limit of observation of the doubling of heat flow is theoretically several million years when the so-called vitrinite-type reactions of low activation energy and low pre-exponential factor are studied. In the basins with an extremely low heating rate this time interval is increased to several ten million years. In these basins there is no significant difference between the intervals that can be studied by reactions of high and low activation enthalpies. In the basins with different features, the time interval to be studied varies between these two extreme values.

We also called attention to the fact that these time intervals which can be studied theoretically, became less wide due to the errors of measurement data. This statement is especially valid in the case where the reaction parameters themselves were determined after the investigation of geological samples.

If one wants to create a thermal history reconstruction of a basin from organic chemical reactions, a large number of measurement data is needed and even in this case the error calculations should be performed since if these calculations are omitted very erroneous results are obtained.

For a more efficient application of organic chemical reactions in the thermal history reconstruction of a basin, the organic geochemists have to determine the activation enthalpy and pre-exponential factors of low-parameter vitrinite-type reactions by means of measurements allowing more insight into more distant times.

As well as this, in order to make the reconstructions more exact, the sensitivity and precision of conversion measurements should be increased. To carry out laboratory experiments, the organic chemical reactions should be selected, which take place in the same way under laboratory conditions (i.e. at higher temperatures) as under geological conditions.

GEOTHERMAL EFFECT OF MAGMATISM AND ITS CONTRIBUTION TO THE MATURATION OF ORGANIC MATTER IN SEDIMENTARY BASINS

HORVÁTH, F.*, P. DÖVÉNYI* and I. LACZÓ**
*Geophysical Department, Eötvös University
Kun Béla tér 2, H-1083 Budapest
** Hungarian Geological Institute
Népstadion út 14, H-1143 Budapest

Abstract

Model calculations were performed to determine the temperature disturbance caused by magmatic intrusions and extrusions. It was our major interest to estimate the influence of this extra heat on the maturation of organic matter in sedimentary basins. The calculations show that there is no real "telemagmatic" thermal effect. Appreciable maturity increase of organic matter is confined to a "zone of influence" which is next to the intrusion and its dimensions are comparable to those of the intrusive body. This moderate influence, however, can be significant if the sedimentary rocks are not very immature but are close to the oil-generation window at the time of volcanic activity. The model calculations were applied to the North Hungarian area of Middle Miocene volcanic activity. We conclude that igneous masses of large dykes and stratovolcanoes could have driven Paleogene sedimentary rocks into the oil-generation window.

Introduction

It is now generally accepted that the generation of natural gases and fluid hydrocarbons is related to the progressive cracking of kerogen, which is a natural constituent of fine grained organic rich sediments (TISSOT & WELTE, 1978). Although several models have been suggested to describe the cracking process, there is little doubt, that it is basically a thermal maturation through geologic time. The most widely used maturity indicator of organic matter is the mean vitrinite reflectance (R_o). This is a very useful parameter because the onset and end of oil generation can be well characterized by given threshold values of vitrinite reflectance. WAPLES (1980) has shown that oil-generation window can be described by the relationship $0.65\ \% \leqslant R_o \leqslant 1.3\ \%$. Furthermore, he has demonstrated that the method originally proposed by LOPATIN (1971) can adequately be used to calculate theoretically the maturity of any sedimentary matter, provided that its temperature history is known.

Temperature changes in the crust of stable continental areas are generally slow and of small magnitude. More rapid and remarkable change of the thermal conditions is associated with tectonically active areas of the Earth. Extensional basins in the Alpine-Mediterranean region, for example, may show 2-fold increase of terrestrial heat flow during an interval of about ten million years (e.g. ROYDEN et al., 1983). The fastest and highest amplitude thermal events are brought about by magmatic activity. Cooling of the high temperature igneous material may significantly increase the temperature of the adjacent sedimentary rocks, thus contributing to their thermal maturation. It is the aim of the present paper to model this process in order to get an idea about the significance of magmatism in hydrocarbon generation. First, we demonstrate that predictions made by conductive modelling are in good agreement with measured vitrinite reflectances in the contact zone of volcanic dykes. Then, it is shown that the increase of maturity is markedly different if the magmatic event occurs in the late rather than in the early stage of basin evolution. The results are applied to North Hungary, where Late Eocene through Middle Miocene subsidence and sedimentation were associated with two distinct periods of volcanic activity.

Mathematical model

Temperature change in infinite homogeneous half space $(x,y,z \geqslant 0)$ due to conductive decay of an initial temperature disturbance can be determined by the integration of the point-source solution (SIMMONS, 1967). If the source is a rectangular prism with $x_1', x_2', y_1', y_2', z_1', z_2'$ coordinates, it is given by the following equation:

$$T(x,y,z \geqslant 0, t) = T_0 + Gz + \int_{x_1'}^{x_2'} \int_{y_1'}^{y_2'} \int_{z_1'}^{z_2'} \frac{T_s}{8(\pi \kappa t)^{3/2}} \left\{ \exp\left[-\frac{(x-x')^2 + (y-y')^2}{4 \kappa t}\right] \right.$$

$$\left. \cdot \left[\exp\left(-\frac{(z-z')^2}{4 \kappa t}\right) - \exp\left(-\frac{(z+z')^2}{4 \kappa t}\right)\right] \right\} dx' dy' dz'$$

where, T_0 = temperature of the surface (z=0)
G = original (undisturbed) temperature gradient in the vertical direction
T_s = initial temperature anomaly of the source
κ = thermal diffusivity of the infinite half space.

If the temperature of the magma at the time of intrusion is T_m, then $T_s = T_m - (T_0 + Gz)$. Integration of this equation can be performed analytically, and we used the formula derived by MUNDRY (1968) for numerical calculations. Sources of more complex shape can be approximated to any required accuracy by a series of rectangular prisms. Moreover, a sequence of volcanic eruptions can also be calculated by simple superimposition of the appropriate solutions.

The main simplifications and some possible improvements of the Mundry's formula are as follows:

i) Omission of latent heat. The latent heat of the magma, which is released during solidification was not taken into account. A precise treatment of the problem is rather complicated, but a good approximation is obtained if one supposes a higher value (T_m^*) for the original temperature of the magma. JAEGER (1964) suggested $T_m^* = T_m + L/c$, where L and c are the latent heat and the specific heat of the magma, respectively.

ii) Homogeneous half space. The formula was derived by assuming that thermal conductivity and diffusivity had the same constant values in the volcanic body and the surrounding rock masses. This may appear a bad assumption, but in fact the average value of these thermal parameters for volcanites and sedimentary rocks are about the same. For example the average thermal diffusivity of volcanites at room temperature is about $8 \cdot 10^{-7}$ m²/s, while this is about $9 \cdot 10^{-7}$ m²/s for claystones and shales (KAPPELMAYER & HAENEL, 1974).

iii) Omission of convective heat transport. Convection of high viscosity magma and hydrothermal processes result in faster cooling of a magmatic mass. This effect can be taken into consideration only by complicated numerical calculations (NORTON & KNIGHT, 1977). However, a fairly good approximation is obtained if only conduction is supposed, but actually the true value of the thermal diffusivity is increased by about 15 % (BUNTEBARTH et al., 1982). It is unreasonable to accept a larger value, because in reality both thermal parameters decrease with increasing temperature for most of the volcanic and sedimentary rocks (KAPPELMAYER & HAENEL, 1974). In all of our model calculations $\kappa = 10^{-6}$ m²/s was used.

As an example, we show in Fig. 1a-d the cooling of a large 25 km x 25 km x 8 km rectangular prism with $T_m = 1000°$ C original temperature in a half space characterized by 10^{-6} m²/s thermal diffusivity and by 33° C/km geothermal gradient background. Calculation was performed for five successive time epochs, but the last one (2.5 Ma) is not shown because there is practically no temperature anomaly available at that time. One can see that even a large plutonic body cools rather fast and marked temperature increase occurs above and close to the body.

Another example is presented in Fig. 2a-b. The cooling of a dyke with 10 m width and 1000° C original temperature was calculated in an infinite half space. Fig. 2a shows the change of temperature with time at different distances from the dyke. Fig. 2b shows, in turn, the temperature profile around the dyke at different time epochs. It can be concluded that the decay of the temperature disturbance is very

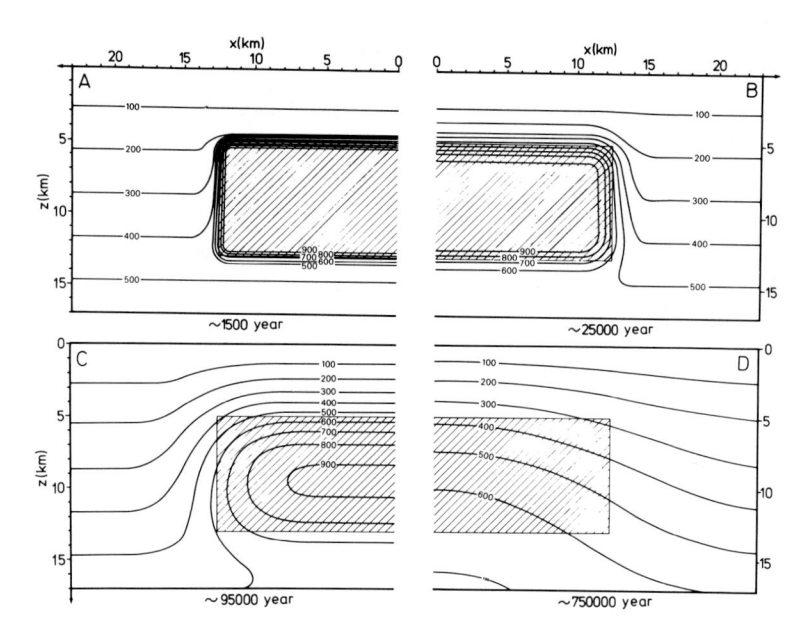

Fig. 1a-d. Temperature field at various times after the intrusion of a plutonic body into a medium characterized by 33 mK/m temperature gradient. The body is a rectangular prism with 8 km x 25 km x 25 km dimensions. The 100° C, 200° C, etc. isolines are shown in a section through the center of the prism and normal to the sidewall of 8 km x 25 km. Because of mirror-symmetry each figure shows only one half of the full section.

fast. High temperatures occur for a period of several months to some years and they are confined to a distance comparable to the width of the dyke.

Calculation of vitrinite reflectance

Maturation of organic matter depends on the temperature history of sedimentary rocks. One of the most widely accepted simple methods for calculation of maturity is that of LOPATIN (1971), as improved by WAPLES (1980). It implies that maturity is related to a Time-Temperature-Index (TTI) which is given by the following formula:

$$TTI = \sum_{n=n_{min}}^{n_{max}} \Delta t_n \cdot 2^n$$

where Δt_n is the time (in million years) the organic matter spent in the n-th 10° C temperature interval. The value of n is shown in Table 1.

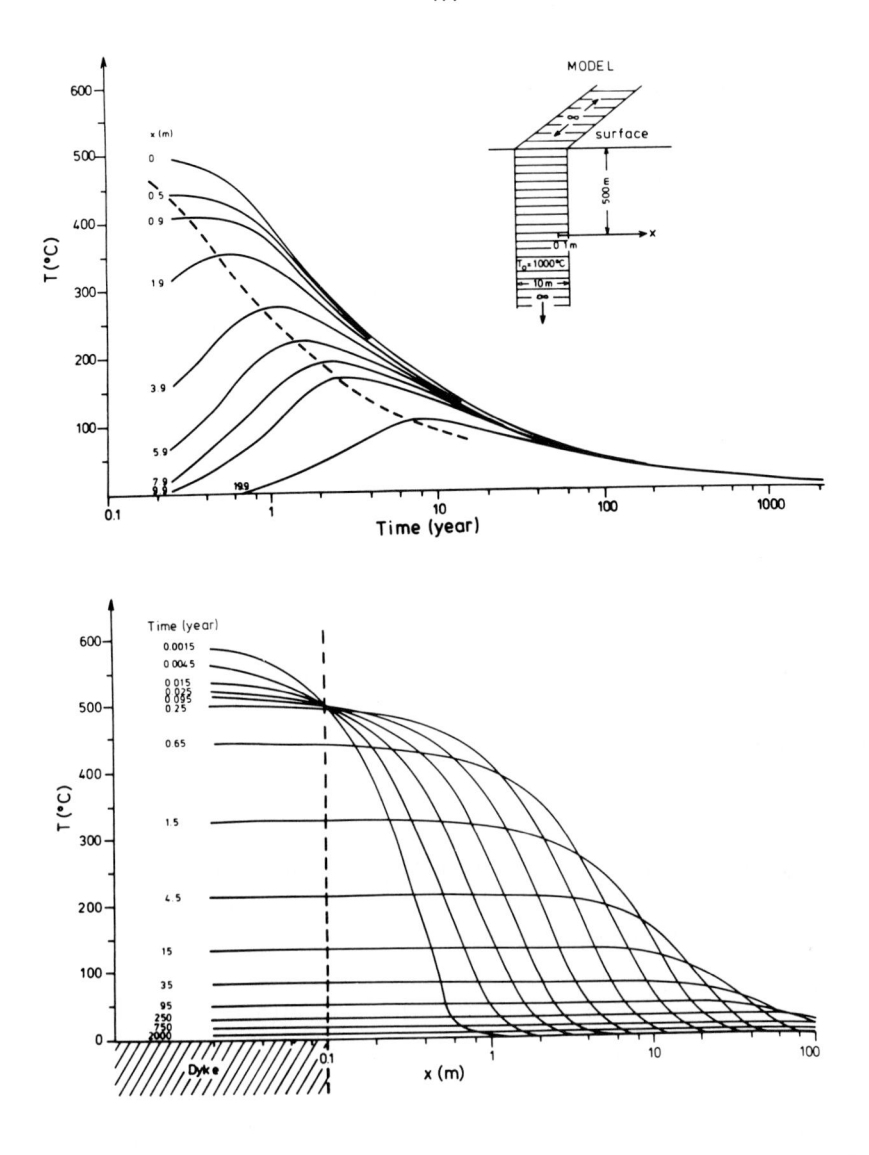

Fig. 2a-b. Temperature disturbance caused by the intrusion of a dyke at t = 0 time epoch, with parameters shown on the inset at the upper right corner. Fig. above gives temperature vs. time curves at different distances from a point within the dyke. The broken line connects the peak of the curves. Fig. below shows temperature vs. distance profiles for various times after the intrusion.

Table 1. Value of temperature coefficient in the LOPATIN (1981) formula

Temperature °C	n	2^n
⋮	⋮	⋮
80- 90	-2	0.25
90-100	-1	0.5
100-110	0	1
110-120	1	2
120-130	2	4
⋮	⋮	⋮

For computation a more convenient form can be used:

$$TTI = \int_0^{t'} 2^{\frac{T(t) - 105}{10}} dt$$

where $T(t)$ is the temperature history of a sedimentary rock in centigrades from the Present ($t = 0$) to the time when the sediment was deposited ($t = t'$). Vitrinite reflectances can be obtained from TTI values with the help of an empirical relationship (Table 4 of WAPLES, 1980) which shows that $\lg R_o = C \lg TTI$, where $C \neq 1$. Accordingly, vitrinite reflectances are not additive quantities. It is only true for the TTI values. In other words, a thermal event, which results in the same increase of TTI, leads to very different reflectance increase depending on the original maturity level of the organic matter. We must be aware of this simple fact when the influence of volcanic heat on the maturation is modelled.

Another point to be considered is related to the very high temperatures, which appear for a while in the proximity of igneous masses (Fig. 1). The Lopatin-Waples formular must not be used for temperatures higher than 200° C, since the TTI-R_o conversion is quite uncertain for high TTI values (KATZ et al., 1982). Therefore another method is needed for the proximal zones to predict the level of maturity. We accepted a method which was originally proposed by BOSTICK (1973) who collected vitrinite reflectance measurements in the contact zone of volcanic dykes. Reflectances are very high at contact and they rapidly decrease with increasing distance. This trend can be explained theoretically by the use of Bostick's diagram, which is shown in the inset of Fig. 3.

This mean vitrinite reflectance vs. temperature diagram was the result of laboratory measurement, when lignitic shale samples were kept at various elevated temperatures and pressures for a period of one month. We have seen (Fig. 2a-b) that elevated temperatures occur for comparable time intervals next to volcanic dykes. Therefore it is

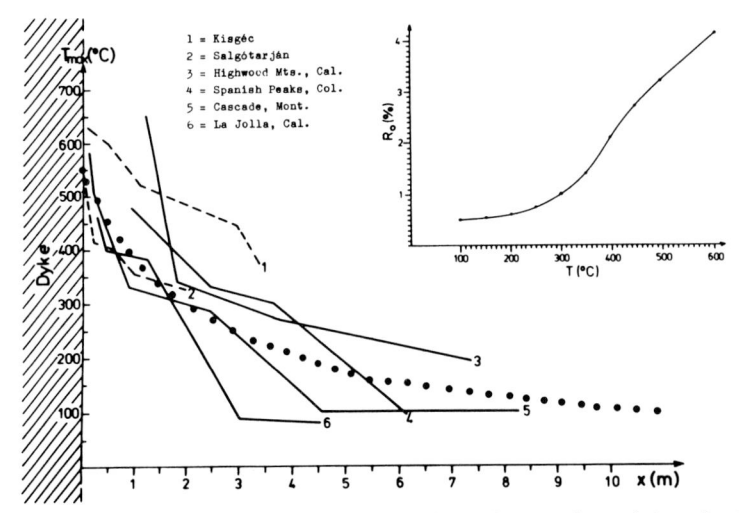

Fig. 3. Maximum temperature vs. distance functions for various dykes in Hungary (1, 2 see location in Fig. 5) and the United States (3 to 6). Temperatures were derived from measured vitrinite reflectances with the help of the Bostick diagram shown in the inset (BOSTICK, 1973). Dotted line shows the same function calculated for a dyke model with 5 m width and 1100° C original temperature.

reasonable, in such cases, to apply the Bostick diagram for conversion of temperatures to vitrinite reflectances or vica versa. It is demonstrated by Fig. 3, which shows that converted temperatures agree fairly well with the calculated ones.

Maturity changes due to "thermal events" in sedimentary basins

In the course of basin evolution sedimentary rocks reach considerable depths and temperatures as the subsidence and burial proceed. The progress of thermal maturation of organic matter can be influenced by thermal events which are brought about by magmatic activity at different times during basin formation. The important role of the timing of these events can be demonstrated by the study of a simple model example.

Let us consider a basin in which a 3 km thick sedimentary complex has accumulated at a constant rate (75 m/Ma) during the last 40 million years. The geothermal gradient has not changed with time and is equal to 33.3 mK/m. This means that temperature increases linearly with depth and now amounts to 110° C at the bottom, if 10° C is supposed at the surface. The present vitrinite-reflectance vs. depth function can be calculated and is shown by the continuous line in Fig. 4. It is now assumed that a volcanic event, 32.2 million years before the present, increased the temperature everywhere in the sediments by 100° C for a period of $0.2 \cdot 10^6$ years. This is the consequence of the fact that 32.2 million years ago the sedimentary column was only 585 m thick and characterized by a maximum temperature of 28° C at the bottom. The thermal event raised the temperature of a cold and immatured sedimentary complex and

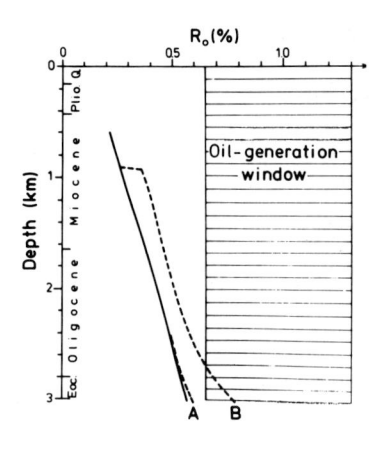

Fig. 4. Model example to show the change of vitrinite reflectance with depth curve, $R_o(z)$, due to thermal events in early and late stage of basin evolution. Solid line gives the $R_o(z)$ curve for a basin characterized by even subsidence and sedimentation (3 km in total, during the last 40 Ma), and constant geothermal gradient (33.3 mK/m). Broken lines show the change of the reflectance curves as a consequence of an overall 100° C temperature increase during the 32.2 - 32.0 Ma and 12.2 - 12.0 Ma periods (A and B events, resp.).

the small increase of reflectivity resulting was overprinted by the later thermal maturation. If, however, the same thermal event had occurred more recently, say 12.2 million years before present, then the available 2085 m sediment would already be moderately matured and much warmer at higher depths. This B event, therefore, resulted in markedly increased vitrinite reflectances and the lower few hundred meters of the section entered into the oil-generation window.

Implication for North Hungary

The model discussed was a simplistic one, but the conditions are fairly similar to the real situation in North Hungary. Here subsidence and marine to lacustrine sedimentation started in the middle Eocene and continued at least up to the middle Miocene, or locally even through the Pliocene. The basin formation was accompanied by two separate phases of calc-alkaline volcanic activity. The first occurred during the late Eocene and the second, a more intensive activity, during the middle Miocene (BALOGH & KŐRÖSSY, 1974). The temperature field is variable in the region, but the average gradient is close to 33.3 mK/m (DÖVÉNYI et al., 1983). The supposition that this value did not change with time seems to be tenable because measured vitrinite reflectances in North Hungary agree well with the predicted values in Fig. 4. Measured values which are apparently not influenced by the volcanic heat are 0.3 % to 0.4 % for Mio-Pliocene, 0.4 % to 0.53 % for Oligocene and 0.53 % to 0.6 % for Late Eocene rocks. Such a basin model is accordingly good enough to use in order to assess the maturity increase caused by igneous masses. We shall not study the first volcanic phase because we have seen that it occurred too early to give notable maturity increase. The second phase was very intensive and produced the stratovolcanic complexes of the Börzsöny and Mátra Mts. (Fig. 5) and possibly large dykes. Two kinds of model are taken into consideration: a stratovolcano and a dyke which broadens downward. The stratovolcano is modelled by a three-dimensional body which is made up from three rectangular prisms as shown in Fig. 6. This was formed 16 million years

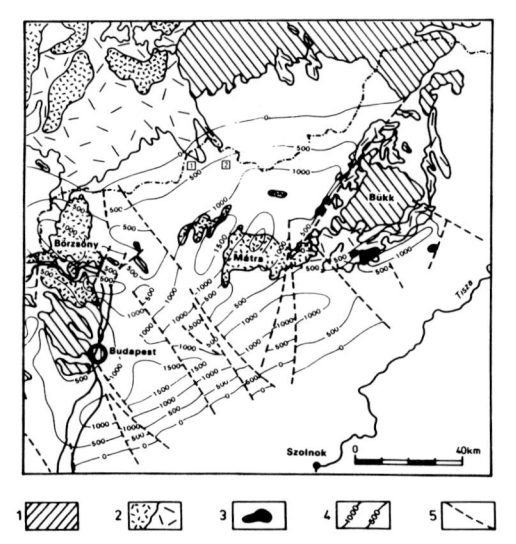

Fig. 5. Simplified geologic map of North Hungary (partly after BALOGH & KŐRŐSSY, 1974 and Map of Oil and Gas Prospection in Hungary, 1978). Heating influence of the Mátra stratovolcano and large subsurface dykes might have contributed to the maturation of Paleogene (dominently Oligocene) source rocks and led to the generation of hydrocarbons which are probably trapped in small fields to the North and East of the Mátra. Legend: 1 = Outcropping pre-Miocene rocks, 2 = Miocene calc-alkaline volcanites, lavas and pyroclastics resp., 3 = Oil and gas fields, 4 = Thickness isoline of Oligocene sediments (in meters). In this area they are below thin to 2 km thick younger sedimentary and/or volcanic cover, 5 = Major faults.

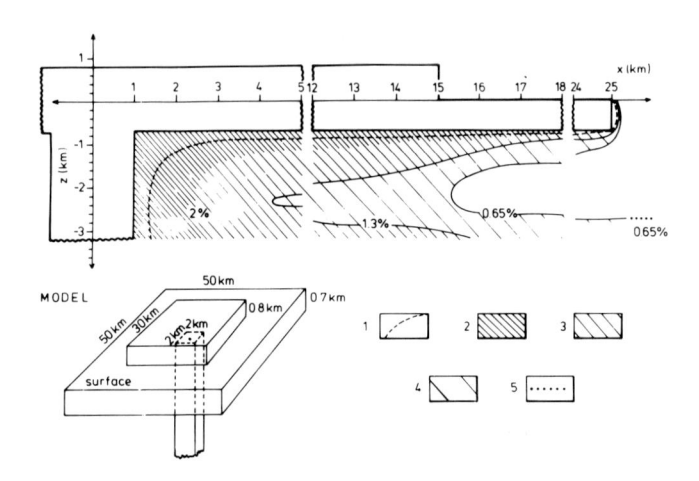

Fig. 6. Increase of maturity of organic matter next to a stratovolcano. The volcano is modelled by three rectangular prisms as shown in the inset. The section goes through the central axis of the body and is normal to the sidewall.
Legend: 1 = R_0 = 2 % vitrinite reflectivity isoline determined by the Bostick diagram, 2 = Extension of the area characterized by $R_0 \geqslant 2$ % reflectance values, as estimated from accepting the Lopatin-Waples method up to 250° C, 3 = Area characterized by 2 % > $R_0 \geqslant 1.3$ % reflectances, 4 = Area characterized by 1.3 % > $R_0 \geqslant 0.65$ % reflectances, 5 = Position of the R_0 = 0.65 % reflectivity horizon in sediments not influenced by the magmatic heat.

ago when the basin was 24 million years old and 2800 m deep. It is supposed that no more subsidence and sedimentation occurred after the volcanic event. 1000° C original temperature was taken for the igneous material. The results are shown in Fig. 6, which gives the present position of lines characterized by 0.65 %, 1.3 % and 2 % vitrinite reflectances. It can be seen that nothing happened far away from the volcano; notable maturity increase is confined to a volume below the thick volcanic strata. The zone of influence is not very large and, in fact, our result should be considered as a maximum estimate, because in reality:

i) formation of a stratovolcano occurs in many cycles during severel hundredthousand years, and

ii) lava flows are alternated by explosion of tuffs, which have a much lesser temperature.

Anyhow, it can be concluded that there is no real "telemagmatic" effect, the zone of influence of an igneous body is about the same volume as the body itself.

The second model we have calculated is a large dyke which broadens downwards (Fig. 7). It is a two-dimensional body with 1000° C original temperature, which intruded the same sedimentary basin 17 million years ago. It can be seen that the zone of influence above the top of the dyke is rather small. It is because the magmatic heat

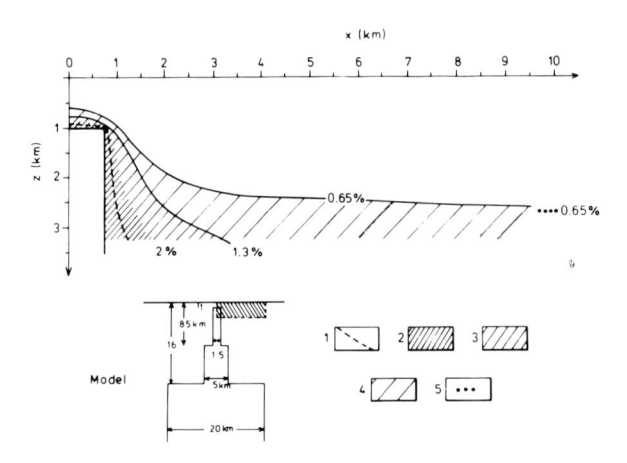

Fig. 7. Increase of maturity of organic matter next to the top of a large dyke as shown in the inset. Legend: 1 = R_O = 2 % vitrinite reflectivity isoline determined by the Bostick diagram, 2 = Extension of the area characterized by $R_O \geqslant 2$ % reflectance values, as estimated from accepting the Lopatin-Waples method up to 250° C, 3 = Area characterized by 2 % > $R_O \geqslant 1.3$ % reflectances, 4 = Area characterized by 1.3 % > $R_O \geqslant 0.65$ % reflectances, 5 = Position of the R_O = 0.65 % reflectivity horizon in sediments not influenced by the magmatic heat.

affected here the cold and immaturated Mio-Pliocene part of the sedimentary succession. The influence is more significant at the sides of the central peak of the dyke, where the Eocene-Oligocene sediments were already warmer and more matured at the time of intrusion. But even here, the zone of influence is not very large. The elevated isoreflectance lines drop down to the normal (undisturbed) level within a few kilometers from the dyke. This limited influence of igneous bodies, however, can be important if potential source rocks would not otherwise reach the main phase of oil generation. This is the case in North Hungary where pelitic Paleogene (dominantly Oligocene) rocks are just at the onset of the oil-generation window. Therefore, we can expect that fluid hydrocarbons could have been generated at places where thick Paleogene sediments are associated with magmatic bodies of Middle Miocene age. Petroleum fields found so far appear to fulfil this prediction, as is shown in Fig. 7.

Generally, we think that buried magmatic complexes in the Pannonian basin - particularly the thick rhyolitic flood tuffs and associated feeder dykes of Middle to Late Miocene age - can be important concerning hydrocarbon prospections. In addition to the fact that magmatic heat contributes to thermal maturation, it should also be taken into consideration that hydrocarbons can migrate and be trapped in volcanoclastics.

Conclusions

The following main conclusions can be drawn from our study:

i) Conductive models combined with the Bostick diagram and the Lopatin method can be used to delineate the influence of magmatism on the maturation of organic matter.

ii) Model calculations show that there is no real "telemagmatic" thermal effect. The zone of influence is confined to sedimentary rocks adjacent to the magmatic body which roughly encompasses the same volume as the body itself.

iii) Maturity increase caused by a given igneous body strongly depends on the pre-existing thermal and maturity conditions in a sedimentary basin. The optimal situation occurs in a not very young basin when fairly matured sediments are driven into the oil-generation window by the heating of igneous material.

iv) This condition probably prevailed at some places in the Pannonian basin during the intensive Middle to Late Miocene calc-alkaline volcanic activity.

Acknowledgements

This work was initiated and supported by the Hungarian Geological Institute. We are particularly grateful to Dr. Á. Jambor for valuable advice and Dr. G. Hámor for permission to publish. We also thank Prof. L. Rybach (ETH, Zürich) for discussion on thermal modelling.

PALEOTEMPERATURES IN THE CENTRAL ALPS, - AN ATTEMPT AT INTERPRETATION

D. WERNER
Institut für Geophysik, ETH Zürich

Abstract

Paleotemperature data contain important information about the uplift history of a mountain range. A simple uplift model for the Gotthard region (Central Alps) is presented. This model satisfies the paleotemperature data but leads to unrealistically high temperatures at greater depths (lower crust, uppermost mantle). To improve the model additional crustal heat sources must be introduced which are assumed to be frictional heat sources caused by crustal overthrusting. This thermal problem also appears in other subregions of the Central Alps. Further problems are related to the thermal transition zone between adjacent crustal blocks with different uplift histories.

Introduction

Paleotemperatures in conjunction with radiometric ages are an important tool to reconstruct the uplift history of a young mountain range like the Alps. The precondition is that rock samples must be available which can tell us about its thermal history. On the other hand, uplift and denudation mirrors the temperature distribution in the earth's crust. This means that the "thermal memory" of such a rock sample contains information about the uplift history of a whole crustal block. This thermal memory is based on a series of different processes, which are the concern of nuclear physics. For all these processes a so-called blocking temperature can be defined, which represents an upper limit of a temperature interval in which nuclear accumulation processes have taken place. The number of the nuclear events, then, is proportional to the time span between present, and the point of time at which the cooling rock sample has passed through the blocking temperature: for instance 500° C for muscovite and phengite Rb-Sr, 350° C for muscovite and phengite K-Ar, 300° C for biotite K-Ar, and 120° C for apatite fission track ages (JAEGER et al., 1967; HUNZIKER, 1974; FREY et al., 1976; PURDY & JAEGER, 1976; WAGNER et al., 1977). From these benchmarks the thermal history of a rock sample during its path towards the

---

*) Contribution no. 483 of the Institut für Geophysik, ETH Zürich

earth's surface can be reconstructed, and a simple formula may be used to estimate uplift rates:

uplift rate = cooling rate/geothermal gradient.

A more precise study, however, cannot be based on this formula, because the geothermal gradient is dependent on depth and time and on the distribution of radiogenic heat sources. In Fig. 1, a simple sketch is shown which may illustrate the thermal changes caused by uplift and erosion.

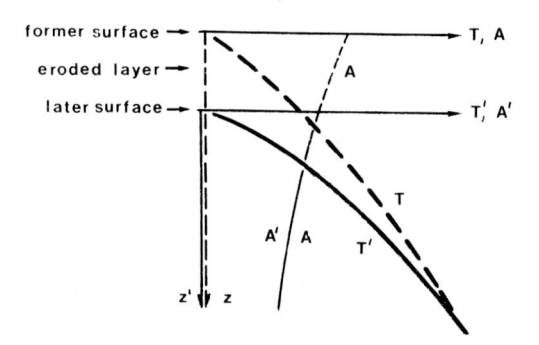

Fig. 1. Sketch of the temperature depth curves before and after a period of uplift and erosion (T and T' resp.). A: radiogenic heat sources.

Several methods have been used to describe the relationship between uplift and temperature field (CLARK & JAEGER, 1969; OXBURGH & TURCOTTE, 1974; ENGLAND, 1978; WERNER, 1980, 1981). In this paper some difficulties related to the interpretation of the paleotemperature data in the Central Alps are discussed. The main problem is to find an explanation for the considerably high paleotemperatures in the crust which does not agree either with near surface geothermal observations, nor with the fact that in the uppermost mantle under the Central Alps relatively low temperatures must be expected (WERNER & KISSLING, 1985).

The problem may be demonstrated by considering the Gotthard region which is a subregion of the Central Alps. Taking a one-dimensional uplift model the geothermal situation in the Alpine lithosphere cannot be interpreted in a satisfactory way. Additional heat sources within the crust must be introduced to fit the paleotemperatures. In this paper a preliminary attempt is made to interprete the high temperature data by assuming that frictional heating due to crustal overthrusting may play a geothermal role. This means that we consider a model which combines one-dimensional uplift, and horizontal shear motions.

It must be noted that at least two long-term processes are neglected here. The first one is related to the size of the region which is built of laterally limited regions (crustal blocks) with different uplift histories and with lateral thermal exchange between adjoining crustal blocks and not of an infinite one-dimensional area. The second process is the crustal thickening during the history of the Alps.

Theoretical remarks

The problem can be expressed by the differential equation describing the heat transport in a moving medium (e.g. LANDAU & LIFSCHITZ, 1978):

(1) $\quad \frac{\partial T}{\partial t} = -\vec{v} \text{ grad } T + \frac{1}{c\rho} \text{ div}(K \text{ grad } T) + A + \frac{1}{2}\eta \left(\frac{\partial v_i}{\partial x_K} + \frac{\partial v_K}{\partial x_i}\right)^2$

where T = temperature, t = time, $\vec{v}$ = velocity of the medium, c = specific heat capacity, ρ = density, K = thermal conductivity, A = strength of radiogenic heat source, η = viscosity, and x_i = cartesian coordinates.

The first term on the right hand side describes the convective portion of the heat transport. The velocity $\vec{v}$ is a given quantity representing a kinematic model of the tectonic history of the Alps, and in writing (1) we have assumed that the medium is incompressible:

(2) $\quad \text{div } \vec{v} = 0$.

The second term describes the conductive heat transport with a temperature dependent conductivity. The heat source A follows the material moving with velocity $\vec{v}$ and satisfies the continuity equation

(3) $\quad \partial A/\partial t = -\vec{v} \text{ grad } A$.

The last term in Eq. (1) describes the frictional heat sources assuming that the medium behaves as a Newtonian fluid.

To implement the boundary conditions we assume that for long-term processes the uplift rate and the erosion rate are equal at all times. This means that the surface remains a level plane with a constant temperature.

In order to demonstrate the thermal problem in the Central Alps by a simple model, we assume that only two processes have taken place: a time dependent uplift and horizontal motions. For the last one a very simple kinematic model may be given by

(4) $\quad v_x(z) = v_0 \left\{ 1 - \frac{2}{\pi} \arctg [c(z-z_0)] \right\}$

where v_0, c and z_0 are constants. This model describes the overthrusting of a crustal layer with constant thickness z_0.

Combining uplift and overthrusting the geothermal model remains one-dimensional and Eq. (1) reduces to

(5) $\quad \frac{\partial T(z,t)}{\partial t} = -v_z(t) \frac{\partial T(z,t)}{\partial z} + \frac{1}{c\rho} \left[\frac{\partial}{\partial z}(K(z,t) \frac{\partial T(z,t)}{\partial z} + A(z,t) + \frac{1}{2} \eta(z)(\frac{\partial v_x(z,t)}{\partial z})^2 \right]$

To evaluate Eq. (5), a simple finite difference method in space and time has been used.

Paleotemperatures in the crust

In the Gotthard subregion of the Central Alps considered here, three fission track data (blocking temperature for apatite: 120° C; ages: 6.7, 7.2, 7.6 m.y.) and two Rb-Sr data (blocking temperature for biotite: 300° C; ages: 15.1 ± 1.8 and 15.7 ± 1.1 m.y.) are known (WAGNER et al., 1977). The errors of the blocking temperatures are in the order of ± 10 %. The fission track ages are topographically reduced values (BUCHLI & WERNER, 1985). From the degree of metamorphism in the Gotthard region follows the original depth of the rock sample which amounts to about 15 km (WAGNER et al., 1977).

The task now is to find a kinematic model such that the thermal history of the rock sample passes through the benchmarks of the paleotemperature data. For this purpose the kinematic field (v_x, v_z) which is the input of the calculation, must be modified to fit the data. Using Eq. (5), the entire temperature variations in space and time have to be calculated. After each time step the burial depth z and the corresponding temperature T_s of the rock sample must be determined.

Fig. 2 shows the uplift history (a), the burial history of the rock sample (b), and its corresponding temperature history T_s (c) passing through the benchmarks of the paleotemperature data. The temperature curve 1 corresponds to a simple one-dimensional uplift model shown in the upper part (a). The original temperature distribution (26 Ma b.p.) to this model is shown as curve 01 in Fig. 3. P1 means the corresponding temperature distribution at present. It is obvious that the curves 01 and P1 in Fig. 3 show unrealistically high temperatures in the lower crust and in the upper mantle which disagrees with the cold body (lithospheric root) to be expected under the Central Alps. This body is characterized by increased seismic velocities (PANZA & MUELLER, 1978; BAER, 1980) and can be interpreted as a negative mantle tem-

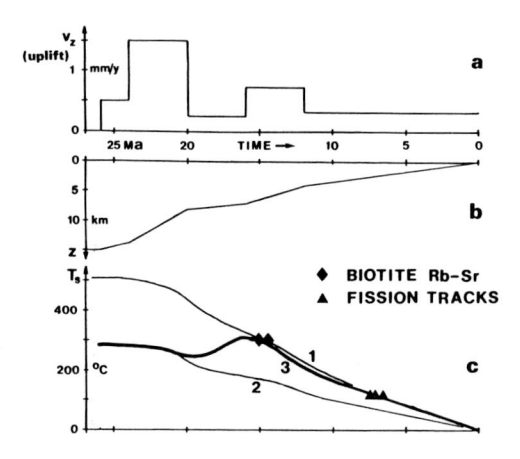

Fig. 2. a: Modelled uplift history of the Gotthard region (Central Alps) which consists of 5 periods of different uplift rates v_z.
b: The corresponding depth changes of a rock sample starting at a depth of 15 km 26 Ma b.p.
c: The corresponding temperature history T_s of the rock sample. Curve 1 represents a simple one-dimensional uplift model leading to unrealistically high temperatures at greater depths (curves O1 and P1 in Fig. 3). Curve 3 corresponds to the same uplift model but with additional heat sources in the crust, whereas curve 2 represents the model without heat sources.

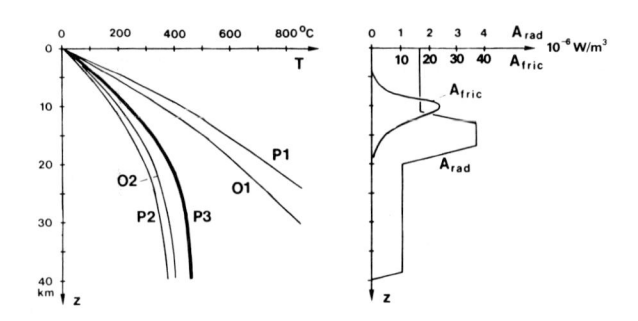

Fig. 3. Left: Original (O) temperature depth curves at the beginning of the uplift history (26 Ma b.p.) and present (P) temperature depth curves. O1 and P1 are unrealistic, based on uplifting only.
Right: Assumed radiogenic heat sources (A_{rad}) at the beginning of the uplift history, and frictional heat sources (A_{fric}) during the time span between 20 and 10 Ma b.p. (note the different scales).

perature anomaly which corresponds to a thermally induced positive density anomaly. Such an anomaly agrees with gravitational observations, on the one hand, and with a possible displacement history of the lithospheric root, on the other. An adequate model (WERNER & KISSLING, 1985) leads to lateral temperature differences up to $-450°$ C (at a depth of about 140 km) related to an undisturbed temperature distribution outside the mantle anomaly.

The curves O1 and P1 in Fig. 3 are not only unrealistic for greater depths but also with respect to near surface geothermal observations. Considering, for instance, a local area within the Gotthard region where extremely high radiogenic heat sources have been observed (Rotondo granite, KISSLING et al., 1978) the surface heat flow amounts to 86 mW/m² (WERNER, 1985). Similar results have been found by BODMER (1982). This means that there is no evidence of remarkable high heat flow values in the Gotthard region, or in the whole of the Central Alps.

A better model can be found by introducing additional heat sources in the crust which are simulated here as frictional heating. In order to fit the benchmarks these sources should have existed between 20 and 10 Ma b.p. That means, the model of Eq. (4) is restricted to a time span of 10 million years. The assumed parameters are: v_o = 4.5 mm/y, z_o = 10 km, c = 0.25. The total horizontal displacement, then, amounts to 45 km. The viscosity at 10 km depth is assumed to be 10^{24} Poise. This simple overthrusting model, however, leads to contradictions from a dynamic point of view: it cannot be explained as a result of gravity sliding processes. Assuming that horizontal motions are only caused by gravity gliding, the viscosity must be remarkably reduced and cannot be in the order of 10^{24} Poise. In this case, the thermal effect of frictional heating can be neglected, and cannot be helpful in interpreting the crustal paleotemperatures in question.

Taking our simple overthrusting model, the curve 3 in Fig. 2c shows the resulting temperatures T_s which satisfies the benchmarks again, but corresponds to more realistic temperatures in the lower lithosphere (curves O2 and O3 in Fig. 3). The same case but without additional crustal heat sources is shown for comparison in Fig. 2c (curve 2) and in Fig. 3 (curve P2). In the framework of one-dimensional modelling the original temperature distributions O1 and O2 are assumed to be steady state distributions which may not be quite realistic for the Alpine region.

The right part of Fig. 3 shows the assumed distributions of the heat sources. The near surface radiogenic heat of the Rotondo granite is extremely high (KISSLING et al., 1978). At the beginning of the uplift history these rocks were situated at a depth of 15 km. The high values (3.77×10^{-6} W/m³) are not considered as representative for the formerly uppermost crustal layers. A value of 1.7×10^{-6} W/m³ has been assumed for the eroded layers. It is obvious that this model of radiogenic heat production must be a speculative one. Another point is unsatisfactory, namely that the one-dimensional uplift model cannot describe the crustal thickening which is connected with the thickening of radiogenic heat sources. Fig. 3 also shows the frictional heat sources from the horizontal motions based on Eq. (4). As mentioned, these heat sources are limited to a time span between 20 and 10 Ma b.p.

Universität Karlsruhe
Geologisches Institut
(Regionale Geologie)
D-7500 Karlsruhe 1, Kaiserstr. 12

Such an interpretation of the paleotemperature data may be questionable but it demonstrates the geothermal problem in the Central Alps. The model shows that paleotemperatures are not only indicators of the uplift history, but also indicators of additional geothermal processes.

Other subregions in the Central Alps

The same thermal problem as in the Gotthard subregion can also be found in other subregions within the Central Alps (Fig. 4, profiles 1 to 5). Most of these regions require a two-dimensional treatment because of their limited size. The degree of metamorphism and the paleotemperature data clearly show that different crustal blocks (subregions) with different uplift histories must be distinguished (WAGNER et al., 1977; WERNER, 1980). A relative vertical movement between adjacent crustal blocks leads to a lateral heat exchange. Therefore the kinematic and geothermal model must be handled at least as a two-dimensional one. An attempt has been made to construct such models along profiles crossing the Central Alps (Fig. 4). The calculations for all the profiles led to the same result as discussed above: additional crustal heat sources must be introduced in order to obtain realistic temperatures at greater depths. The resulting uplift histories are shown in Fig. 5 (BUCHLI & WERNER, 1985).

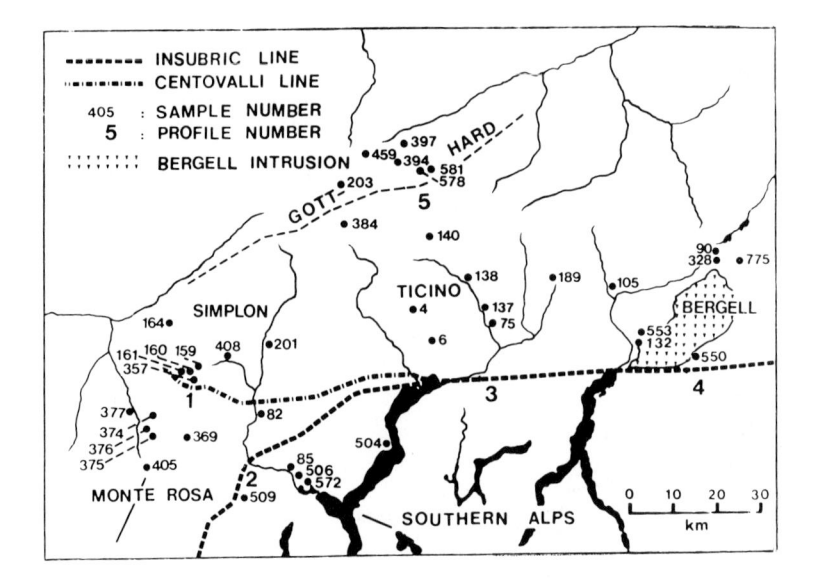

Fig. 4. Location map of the Central and Southern Alps with different profiles for different uplift histories (BUCHLI & WERNER, 1985)

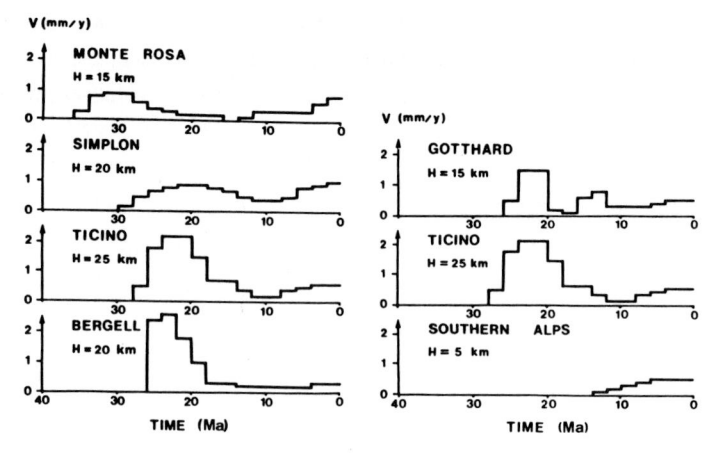

Fig. 5. Modelled uplift histories for subregions from west to east (left), and from north to south (right) after BUCHLI & WERNER (1985). H means the total uplift. For location see Fig. 4.

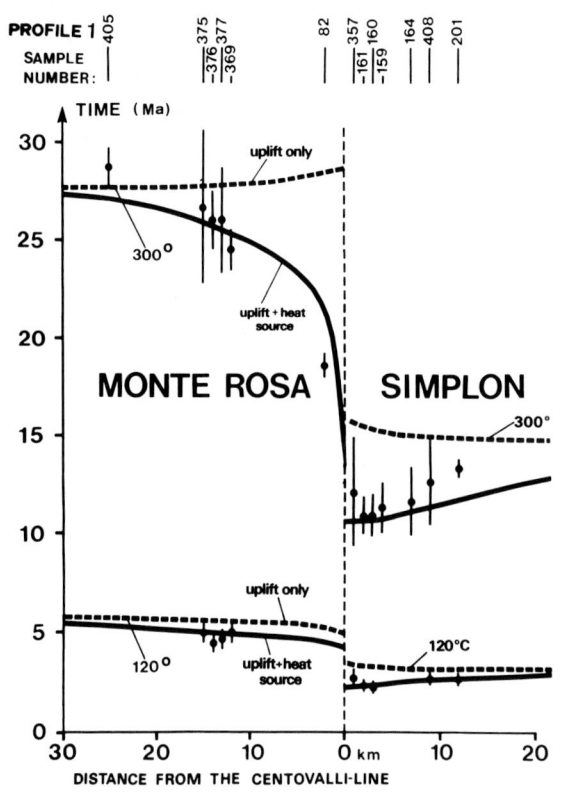

Fig. 6. Cooling history for two crustal blocks, Monte Rosa and Simplon (see profile 1 in Fig. 4) with different uplift histories. The point of time at which the rock samples should pass the blocking temperatures of 300° C and 120° C are indicated by the broken lines. Times at which rock samples (♦) passed the blocking temperatures 300° C and 120° C resp. are also indicated. To obtain a better data fit, frictional heating between the crustal blocks has been introduced, indicated by the solid lines (BUCHLI & WERNER, 1985).

A particular result of these calculations is shown in Fig. 6 for the transition zone between the adjacent crustal blocks Monte Rosa and Simplon, which are separated by a deep-reaching tectonic lineament (Centovalli line). It should be expected that paleotemperature data from locations near the Centovalli line reflect the thermal transition zone. In order to study the thermal situation within this zone, four parameters must be taken in consideration: 1. time, 2. depth of rock samples, 3. temperature of rock samples, 4. distance of rock samples from the Centovalli line. In Fig. 6 only three parameters can be seen: time, temperature, represented by modelled lines for 300° C and 120° C, and distance between the idealized boundary of two blocks (Centovalli line) and the locations where the rock samples come from. Furthermore, it is assumed that the two uplift histories are valid for each block as a whole, excepting a narrow frictional zone (width of the Centovalli fault in the order of 100 m). The interesting stage now, is to compare the broken lines in Fig. 6, with respect to the 300° C paleotemperature data. The broken lines represent a calculation result which is based only on thermally conductive contact between the two crustal blocks. Taking, for instance, the point of time 29 Ma b.p. at which the uplift of the Monte Rosa block was greater than the uplift of the Simplon block (see Fig. 5), we must except higher temperatures in the Monte Rosa than in the Simplon. The lower temperatures in the Simplon should influence the near Simplon parts of the Monte Rosa. This means that a near Simplon rock sample of the Monte Rosa block should pass the 300° C temperature point earlier than a "typical" rock sample from the Monte Rosa far from the Simplon. The data, however, show a reversed tendency. Rock samples from locations near the Centovalli line reach the blocking temperature later than expected from the conductive thermal model, and it seems that the frictional zone between the blocks acts as a heat source (BUCHLI & WERNER, 1985). A similar behaviour can also be found in transition zones of other crustal blocks within the Central Alps, for instance between the Ticino block and the Southern Alps. Frictional heating due to different vertical motion may be a possible explanation of this phenomena.

In all cases, the paleotemperature data in the Central Alps contain not only information for the different uplift histories, but offer new problems concerning the relationship between tectonics and geothermics.

GEOTHERMAL STUDIES IN OIL FIELD DISTRICTS OF NORTH CHINA

WANG JI-AN, WANG JI-YANG, YAN SHU-ZHEN and LU XIU-WEN
Institute of Geology, Academia Sinica
P.O.Box 634, Beijing, China

Abstract

In North China, Tertiary sediments give the main oil-genetic series. The mean value of terrestrial heat flow density has been considered to be 60 - 65 mW/m², and the geothermal gradient in Tertiary sediments usually ranges from 30 to 40° C/km in the region studied. Supposing that the onset of oil generation lies at about 90° C, the upper limit of the depth of oil-generation is at about 2000 to 2500 m depth. Recent paleogeothermal studies using vitrinite reflectance, clay and authigenic minerals, as well as other methods showed that in Eocene the geothermal gradient has been higher than at present. Some results were obtained and discussed.

Introduction

North China is rich in oil resources and recently became one of the main resource-bases for energy supply in China. In some large-scale Mesozoic-Tertiary sedimentary basins such as Lower Liaohe, Central Hebei, Northern Shandong, oil-gas deposits of commercial interest have been found in many places. The North China Plain (including Lower Liaohe) geologically is a large-scale Meso-Cenozoic basin of faulting-depression origin which developed on the Pre-Cambrian basement (ZHANG WEN-YOU et al., 1982). The structural framework of the region has been formed during the "Yintze" tectonic cycle (225 to 195 m.y.). Pre-Cambrian rocks are widely exposed in the mountainous area surrounding the Plain, and Paleozoic sedimentary strata of platform type as well as Mesozoic clastic and volcanic rocks of terrestrial origin are intermittently distributed in the periphery of the Plain. Within the Plain, on a series of rises and depressions of Pre-Cenozoic strata, Cenozoic sediments are lying. The block-faulting movements initiated since Mesozoic have also been continued in the Cenozoic. In spite of some differences in Cenozoic sedimentation history, in the early Eocene (Sahejie, especially early Sahejie), the North China Plain as a whole subsided rapidly. At Oligocene (Dongying), the amplitude of subsidence decreased gradually, but in some places such as Lower Liaohe and partly Shandong, the thickness of sediments has still been recorded as thick as 1000 m or even more. By the end of early Tertiary, the Plain had come to the end of intensive subsidence, and hence, a thickness of several hundred meters for the Miocene (Guantao) and Pliocene (Minghuazhen) sediments are usually observed in most areas with the exception of the Bohai Bay area, where the Pliocene sediments were found as thick as 1000 m or more. Since Quaternary, the North China Plain has gradually stepped into a stage of peneplaination development; the thickness of Quaternary sediments is less than 400 m.

The vertical crustal movement, dominated in Cenozoic era and accompanied by the block-faulting processes, has been regarded as the main form of tectonic activity in this area. Meanwhile, folding process developed weakly, Early Tertiary basaltic volcanism, characterized by fissure eruption and associated with sediments, was extensive in the northern part of the North China Plain. Since late Tertiary, the magnitude and intensity of basaltic volcanism decreased significantly and the Quaternary volcanism has only been found in some local places.

In the North China Plain there exists two types of oil-gas deposits of commercial interest:

1. The so-called "buried hill" type of deposits with mainly Pre-Cenozoic reservoirs of early Paleozoic to Sinian (late Proterozoic) carbonate rocks.
2. Oil-gas deposits with Cenozoic reservoirs of various coarse clastic rocks.

The total thickness of Cenozoic sediments in some strongly subsided depressions amounts to 7000 - 8000 m or even more, and in most areas a thickness of 3000 to 4000 m still remains. Apparently, the Cenozoic sediments, especially the early Tertiary ones, have been considered to be the main oil-genetic series because of the more favourable geological and geothermal settings: the abundant source of organic material; the appropriate temperature conditions for petroleum maturation; and the relatively stable tectonic environment.

Main geothermal features

Recent studies indicate that the oil field district of North China is characterized by a relatively high geothermal setting. The mean heat flow density value of the North China Plain has been thought to be 63 mW/m^2 with the individual values ranging from 61 to 74 mW/m^2 (Geothermal Res. Div., 1979a,1979b; WANG et al., 1981; DENG & WANG, 1982) (Fig. 1). In many oil fields of North China, a higher geothermal gradient has been observed in comparison with other oil fields in China (Fig. 2). Based on numerous temperature logs, a temperature map for the depth of 2000 m (Fig. 3), and a map of geothermal gradient for the whole Cenozoic sediments (Fig. 4) were recently compiled. The temperature at a depth of 2000 m is usually about 75 to 85° C. A geothermal gradient of 30 to 40° C/km covers most (70 %) of the area with higher values in the northern Shandong to the South-East of Bohai, and with lower values at the piedmonts.

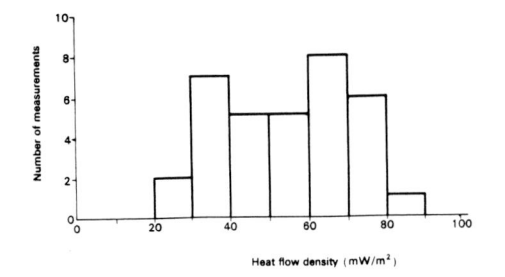

Fig. 1. Histogram of heat flow densities in North China

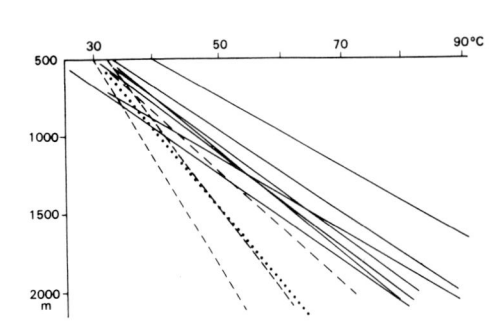

Fig. 2. Geothermes for some oil fields in China. Solid lines: North China; dashed: Sichuan Basin; dotted: Shanxi Basin

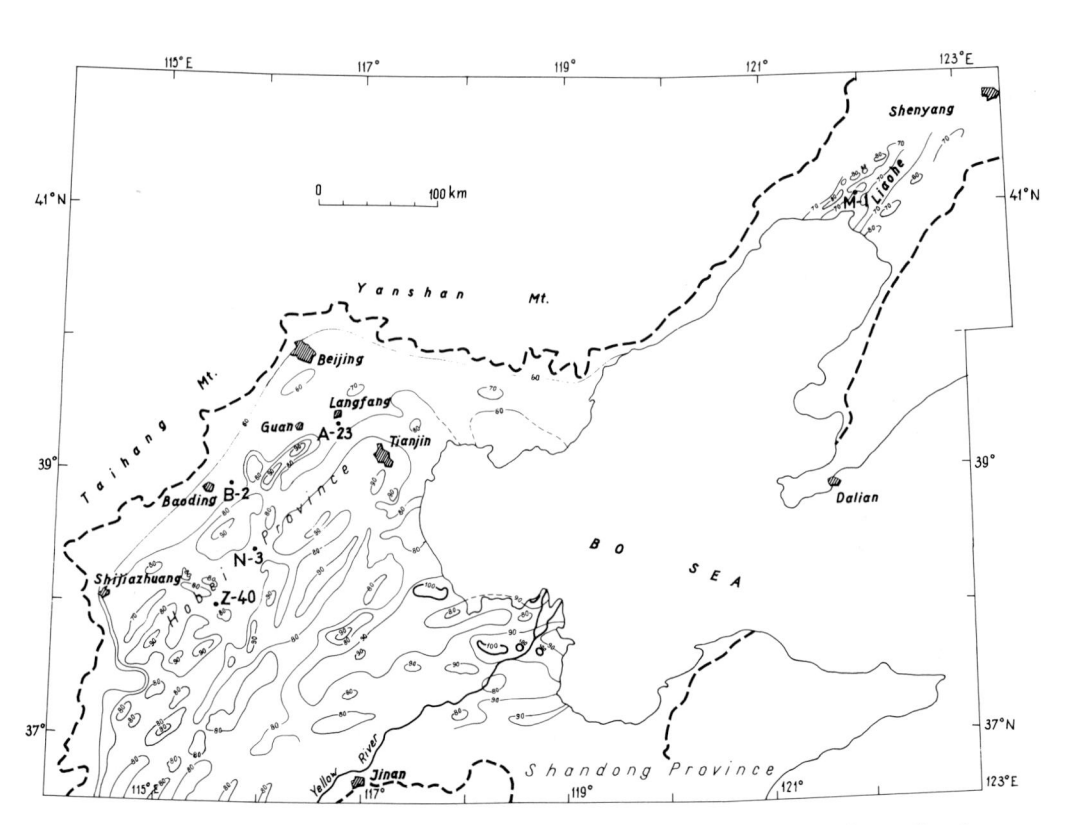

Fig. 3. Geotemperatures in the depth of 2000 m in the northern part of the North China Plain and the Lower Liaohe area

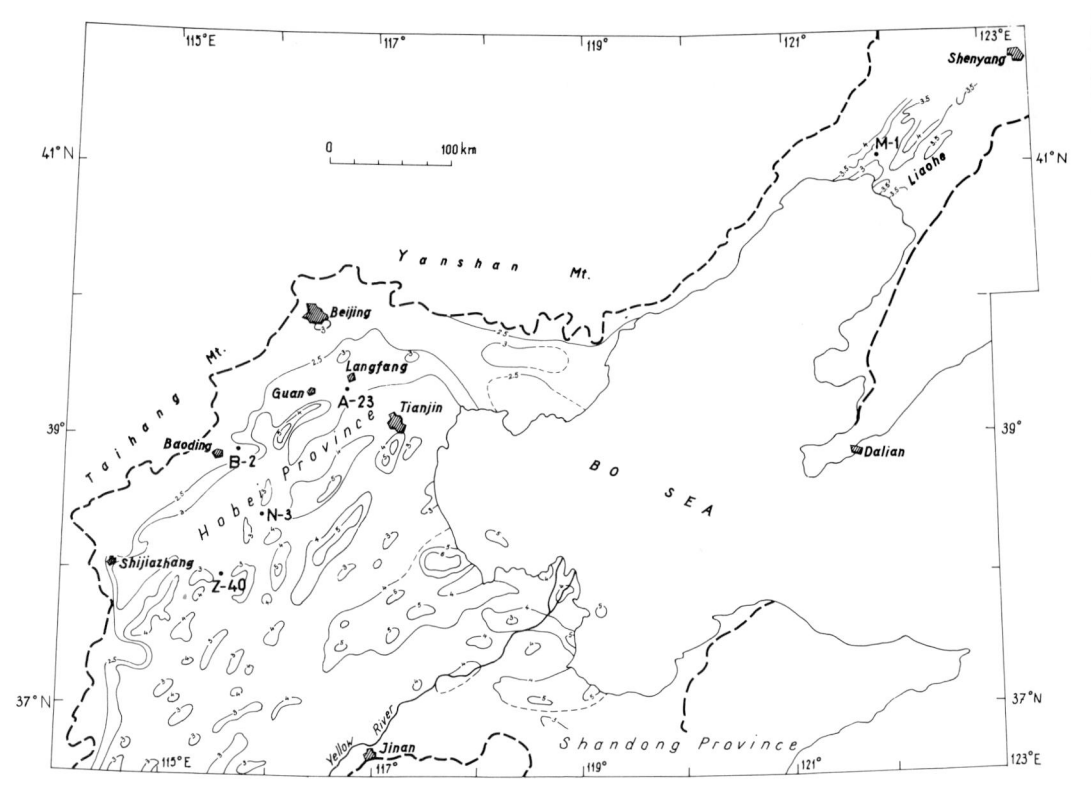

Fig. 4. Geothermal gradients (°C/100 m) in the Cenozoic sediments of the northern part of the North China Plain and Lower Liaohe area

The relatively high temperature of the Cenozoic strata in the North China Plain is favourable for the maturation of hydrocarbon material in rather shallow depths, and also for the occurrence of petroleum.

The measured heat flow density in mountainous area surrounding the North China Plain is about 46 mW/m², and the geothermal gradient is of 15 to 20° C/km, which would be thought to be the typical values for an old stable tectonic unit.

The geothermal regime is controlled by some geological factors, among which the topography of the basement rock and the hydrogeological setting are considered to be the most important ones in the oil field district of North China (Geothermal Res. Div., 1978,1980). As Figs. 3 and 4 show the pattern of isotherms and geothermal gradients are consistent with regional NE to NNE trend. The high temperature zones are correlated to the rises, and the low temperatures to the depressions. This correlation can be explained by the high conductivity of older rocks, as was verified by model calculation using finite element techniques (XIONG LIANG-PING & GAO WEI-AN, 1982; XIONG LIANG-PING et al., 1983). On Fig. 5, the geothermal gradients in Cenozoic

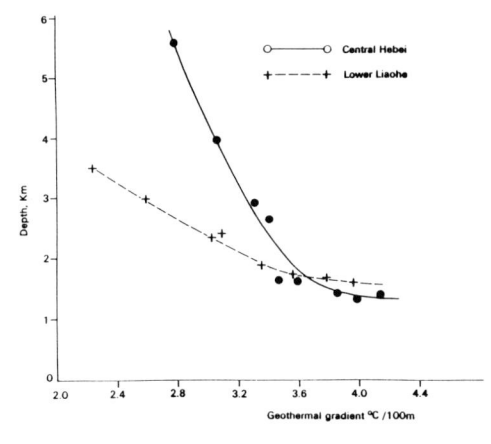

Fig. 5. Geothermal gradients in Cenozic sedimentary cover versus the depth of Pre-Cenozoic and Sinian basement, for two areas in the North China Plain

sediments versus the depth of Pre-Cenozoic or Sinian basement rocks are plotted for two areas, demonstrating this correlation.

Another controlling factor is the hydrogeological setting. The low temperature and low gradient zones at the piedmonts of Yanshan and Taihang Mts. are caused by the strong circulation of the ground water at shallow depths. Along faults or fracture zones, abnormal geothermal gradients and temperatures occur, having resulted from the upward directed water circulation. Observations indicate that the abnormal gradient may reach a value of greater than 50 to 60° C/km which would be regarded as a criteria to identify the existence of a geothermal anomaly caused by convection. In Liaohe Basin, the geothermal gradient for the late Tertiary sediments (usually less than 25° C/km) seems to be always less than that of the early Tertiary series (35 to 40° C/km). Apparently, it resulted from the strong water flow in the aquifers of late Tertiary sediments.

Paleogeothermics

The Cenozoic sediments of North China Plain are considered to be the main oil-genetic series (WANG, 1981; WANG et al., 1983). The thermal history during the Cenozoic era were studied by means of the following two approaches:

1. Analysing in detail the history of sedimentation and erosion, the thermal history for certain oil-genetic series was reconstructed, supposing that both heat flow and geothermal gradient in the past and at present are the same;

2. Using geothermometers such as vitrinite reflectance, clay and authigenic minerals, the maximum temperature or temperature range, which the layer studied has undergone during Cenozoic time, could be determined.

In Fig. 6, the values of vitrinite reflectance versus depth for some wells in the oil field districts of North China are plotted. For a well labelled M-1, located in the

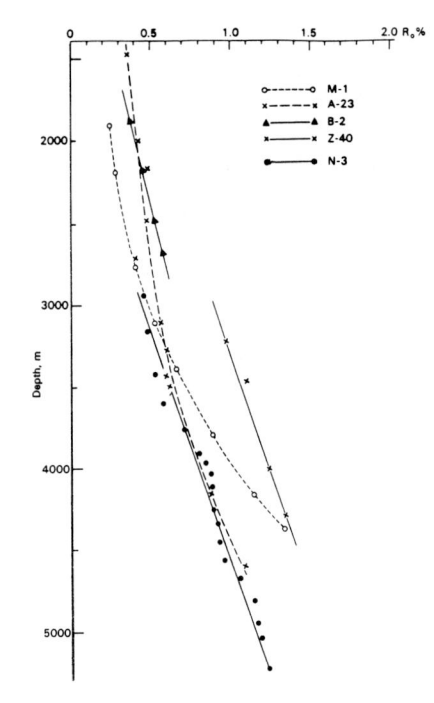

Fig. 6. Variation of vitrinite reflectance with depth in wells of oil fields (M-1 etc.)

western part of Liaohe Oil Field, the results obtained are as follows:

1. Coalification gradient increases gradually from 0.025 R_o %/100 m to 0.065 during the period of mid Eocene to early Eocene, which is consistent with observations in many other oil fields of the world (CASTAUR & SPARKS, 1974; HACQUEBARD, 1975).

2. Relatively low vitrinite reflectances (R_o) correspond to high temperatures, for instance, R_o = 0.4 % corresponds to 95° C and R_o = 1.3 % to 152° C. It is somewhat different with results reported for other oil fields in the world (KARWEIL, 1975). It is supposed that the type and the younger age of organic material in the sediments of Liaohe Oil Field is responsible for this discrepancy.

3. The relationship between vitrinite reflectance and present (probable maximum) temperature for the Liaohe Oil Field can be expressed by:

(1) $R_o \% = 0.00000285 \, T^{2.591}$ (T in °C)

Another example was taken from a well labelled N-3, situated in the Central Hebei Oil Field (Fig. 6). Some similar results were obtained:

1. During the period from late to early stage of early Tertiary, the coalification gradient exhibits a slight value: 0.04 R_o %/100 m.

2. Similarly to the Liaohe Oil Field, the present temperatures of 105 and 158° C correspond to 0.5 to 1.25 R_o %, respectively.

3. A roughly linear relationship exists between R_o and T:

(2) $\quad R_o \% = T/80.9494 - 0.745$ (T in °C)

It must be pointed out that equations (1) and (2) are a result of a second power equalisation of the curves M-1 and N-3 in Fig. 6. The sediments in these two wells are deposited continuously and no erosion existed. Therefore the temperature at a certain depth may be regarded as the maximum temperature during geological time.

More or less similar relationships between R_o and T were also obtained in other oil fields in North China.

For paleogeothermal analysis on a regional scale, the following two approaches were used:

1. Using equations (1) and (2), a reconstruction of the paleotemperature regime has been carried out for the regions where the R_o, sedimentary and present geothermal data suffice and are good enough. Afterwards, we have compared the result obtained with the results calculated by the methods of several investigators (LOPATIN, 1971; WAPLES, 1980; KARWEIL, 1975; HOOD et al., 1975). The results agreed more or less satisfactorily. For example, the calculated R_o at 4300 m depth in well M-1 by Lopatin's method is of 1.26 %, while the measured value is of 1.27 %. The calculated R_o value of Karweil's approach is lower, only 1.1 %, using the present geothermal gradient (33° C/km) for calculation. The calculated paleotemperature and paleogeothermal gradient appears to be still higher by Hood's LOM graph.

2. For a region where R_o data are not sufficient to do this, an estimation was used based on a geological comparison with similar tectonic units and sedimentation history.

The results thus obtained are shown in Table 1. It should be emphasized that the paleogeothermal gradients presented in Table 1 are averaged values since early Tertiary, whereas the present-day's gradients are the mean values for the whole Cenozoic group. Nevertheless, the data set of Table 1 suggests that a tendency of decreasing geothermal gradient with time existed during Cenozoic period, consistently with the evolution of this area (WANG et al., 1983).

Table 1. Mean present and Tertiary geothermal gradients in North China

Locality		Tertiary geothermal gradient (°C/km)	Present geothermal gradient (°C/km)
Central Hebei	Raoyang Sub-depression	28	28
	Langfang-Guan Sub-depression	29 - 31.5	27 - 28
	Baoding Sub-depression	41	27
	Shulu Sub-depression	36 - 38	26 - 28.5
	Wuji Sub-depression	50 (M_z)	40
Liaohe	Northern part of the East Sub-depression	33 - 34	30 - 31
	Mid part of the East Sub-depression	33 - 43	30 - 31
	Western slope of the East Sub-depression	30	31
	Southern part of the Central Sub-uplift	34	33
	Mid part of the West Sub-depression	33	32
	Southern part of the West Sub-depression	34	30
	Western slope of the West Sub-depression	38	27
Northwestern Shandong	Huimin Sub-depression	36	36

Geothermics and hydrocarbon resources

The 90° C and 150° C and R_o = 0.5 and 1.3 % are taken as the lower and upper limits of the generation of liquid petroleum ("liquid window"), as suggested by others (CONNAN, 1974; HACQUEBARD, 1977; HOOD et al., 1975; TISSOT & WELTE, 1978; PHILIPPI, 1965).

As seen on Fig. 4, the isotherm of 90° C is located at a depth of about 2000 to 2100 m in northwestern Shandong, and 2300 to 2400 m in the Liaohe Basin, as well as between these two depths in Central Hebei. At piedmonts of Taihang and Yanshan Mts., and in some deep sub-depressions such as Baoding, Langfang-Guan in Central Hebei, and several other ones in Liaohe, the 90° C isotherm is situated at a depth of 2500 to 2700 m, that is, much deeper. It is clear that almost all of the Cenozoic group in the North China Plain are characterized by high oil-genetic potential, among which the Sahejie and the Dongying formations are considered to be the best because of their abundant organic material and high ability to hydrocarbon discharge as well as the suitable temperature conditions.

The area favourable for a high degree of kerogen maturation Sahejie formation, however, is confined to the north of the Shijiazhuang-Jinan line with a moderate depth of the buried basement rock. In the area to the south of Shijiazhuang-Jinan and on uplifts, the oil-genetic potential of the Sahejie formation is very low owing to the small sedimentation during the Eocene period, or to the erosion afterwards. The area favourable for the oil-genetic process in the Dongying formation is rather limited, only to some deep sub-depressions of Central Hebei and northwestern Shandong, as well as the local part of the Liaohe Basin. The above oil-generating area suggested by geothermics is in good agreement with the real distribution of oil deposits in North China.

In Fig. 7 the thermal history at the base of the Sahejie and Dongying formations is presented. It is obvious that for the Sahejie formation the threshold temperatures were reached before early Tertiary only at places of strong subsidence. By the end of Miocene and up to Pliocene, the temperatures of "liquid window" seem to be reached in most areas. For the Dongying formation, the threshold temperature was partly reached only after Pliocene. In most areas, the present-day's temperature is still beyond the threshold value.

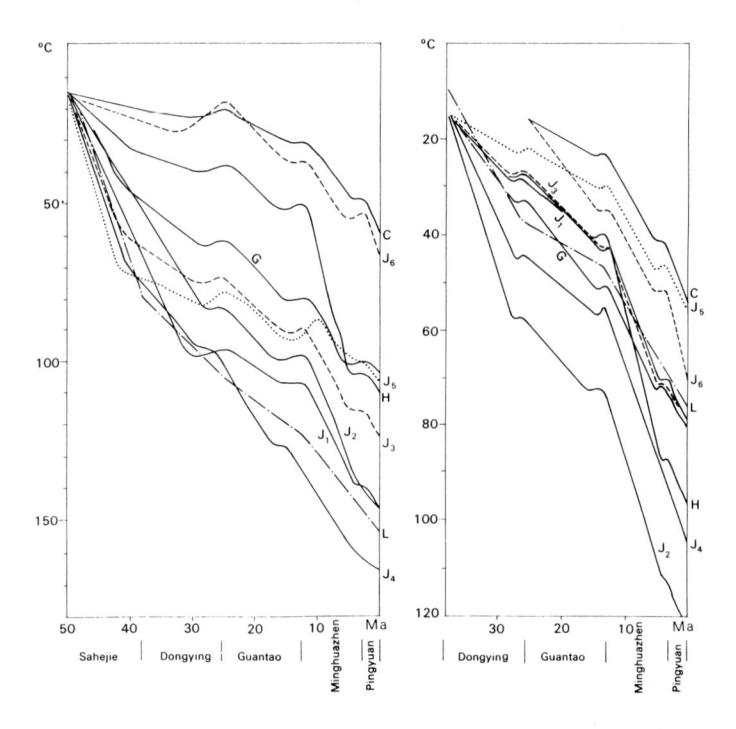

Fig. 7. Temperature history of the bottom of Sahejie (left) and of Dongying formation (right). J - Central Hebei Depression, Sub-depressions: J_1 - Wuging, J_2, J_3 - Baxian, J_4 - Raoyang, J_5 - Baoding, Sub-uplift: J_6 - Ningjin, H - Huanghua Depression, Quikou Sub-depression, G - Yiyang Depression, Cheshen Sub-depression, L - Western Liaohe Depression, C - Cangxian Uplift

Conclusions

1. By the end of early Tertiary, the Sahejie formation in most depressions has come into a stage of the onset of oil-generation, and in Pliocene, a stage of the end of oil-generation has been reached.

2. To date, the Dongying oil-generation series (Oligocene in age) reached the end of oil generation only in a very small amount even in some deep depressions.

3. Guantao (Miocene in age), Minghuazhen (Miocene to Pliocene) formations are not basically oil-generating series, as both the organic material content and temperature conditions were unfavourable.

4. On uplifts, all the oil deposits found in Sahejie, Dongying and other series are believed to be the results of secondary migration.

5. According to the temperature values, deep depressions are the most promising areas for oil and gas prospecting in North China, because of the thickness of oil-generating series as well as of the relatively low geothermal gradient (usually $<30°$ C/km), which makes the depth interval of oil-generation very large. In some deep depressions such as Liaohe, Langfang-Guan, Baoding, the lower limit of the liquid window sometimes lies at 4500 m and even more.

REFERENCES

ABBOTT, G.D., C.A. LEWIS & J.R. MAXWELL (1984): Laboratory simulation studies of steroid aromatisation and alkane isomerisation, in: P.A. SCHENCK, J.W. de LEEUW & G.W.M. LIJMBACH (eds.), Advances in Organic Geochemistry, 31-38 (Pergamon Press)

ABBOTT, G.D., C.A. LEWIS & J.R. MAXWELL (1985): The kinetics of specific organic reactions in the zone of catagenesis, Proc. Roy. Soc. London, Series A (in press)

ABELSON, P.H. (1967): Conversion of biochemicals to kerogen and n-paraffins, in: P.H. ABELSON (ed.), Researches in Geochemistry, Vol. 2, 63-86 (John Wiley & Sons)

ADLER, R. (1961): Der "Westerholter Block", eine tektonische Einheit des Ruhrkarbons, Bergbauwiss. $\underline{8}$, 428-430

AGIP (1977): Temperature sotterane, 1390 p., Milano

AHRENDT, H., J.C. HUNZIKER & K. WEBER (1978): K/Ar-Altersbestimmungen an schwachmetamorphen Gesteinen des Rheinischen Schiefergebirges, Z. dt. Geol. Ges. $\underline{129}$, 229-247

AMMOSOV, I.I., B.G. BABASHKIN & L.S. SHARKOVA (1975): Bituminit nizhnekembriyskikh otlozheniy Irkuts-kay neftegazonosnoy oblasti, in: I.V. YEREMIN (ed.), Paleotemperatury zon Nefteobrazovaniya, 25-59, Moscow (Nauka Press)

AMMOSOV, I.I., V.I. GORSHKOV, N.P. GRECHISHNIKOV & G.S. KALMYKOV (1977): Paleogeotermicheskiye kriteriyi razmesbeniya neftyanykh zalezhey (Paleogeothermic criteria of the location of petroleum deposits), Leningrad (Nedra Press)

AOYAGI, K. & T. ASAKAWA (1977): Primary migration of petroleum during diagenesis, J. Jap. Ass. Petrol. Technol. $\underline{42}$, 76-89 (in Japanese)

AOYAGI, K. & T. ASAKAWA (1984): Paleotemperature analysis by authigenic minerals and its application to petroleum exploration, Am. Ass. Petrol. Geol. Bull. $\underline{68}$, 903-913

AOYAGI, K. & T. KAZAMA (1977): Diagenetic transformation of minerals in the Cretaceous and Tertiary argillaceous rocks of Japan, Prof. Kazuo Huzioka Memorial Volume, Mining College of Akita Univ., Akita, Japan, 151-160 (in Japanese)

AOYAGI, K., N. KOBAYASHI & T. KAZAMA (1975): Clay mineral facies in argillaceous rocks of Japan and their sedimentary petrological meanings, Proc. Intern. Clay Conf., Mexico City, 101-110

ARNASON, B. (1976): The hydrogen-water isotope geothermometer applied to geothermal areas in Iceland, Geothermics $\underline{5}$, 75-80

ARNORSSON, S. (1975): Application of the silica geothermometer in low-temperature hydrothermal areas in Iceland, Am. Jour. Sci. $\underline{275}$, 763-785

BACHMANN, G.H. & M. MÜLLER (1981): Geologie der Tiefbohrung Vorderriss 1 (Kalkalpen, Bayern), Geol. Bavarica $\underline{81}$, 17-53

BAER, M. (1980): Relative travel time residuals for teleseismic events at the new Swiss seismic station network, Ann. Géophys. 36/2, 119-126

BAKER, E.W. & J.W. LOUDA (1983): Thermal aspects in chlorophyll geochemistry, in: M. BJORØY et al. (eds.), Advances in Organic Geochemistry (1981), 401-421 (Wiley)

BALOGH, K. & L. KŐRŐSSY (1974): Hungarian Mid-Mts. and adjacent areas, in: M. MAHEL (ed.), Tectonics of the Carpathian Balkan Regions, Geol. Inst. D. Stur, Bratislava/UNESCO, 391-403

BARBER, P.M. (1982): Palaeotectonic evolution and hydrocarbon genesis of the central Exmouth Plateau, Austral. Petrol. Explor. Assoc. J. 22, 131-144

BARKER, C.E. (1979): Vitrinite reflectance geothermometry in the Cerro Prieto geothermal field, Baja California, Mexico, Inst. of Geophysics and Planetary Physics, Univ. of California, Riverside, UCR/IGPP Report 79/10, 105 p.

BARKER, C.E. (1979): Organic geochemistry in petroleum exploration, Amer. Assoc. of Petrol. Geologists, Course Note Series, No. 10, 159 p.

BARKER, C.E. (1983): Influence of time on metamorphism of sedimentary organic matter in liquid-dominated geothermal systems, western North America, Geology 11, 384-388

BARKER, C.E. (1984): Influence of time on metamorphism of sedimentary organic matter in liquid-dominated geothermal systems, western North America (Reply), Geology 12, 690-691

BARKER, C.E. (in press): Comparison of first order kinetic models of kerogen thermal maturation, in: N.D. NAESER & T.H. McCULLOH (eds.), Assessment of the Thermal History of Basins, Society of Economic Paleontologists and Mineralogists, Special Publication

BARKER, C.E., B.L. CRYSDALE & M.J. PAWLEWICZ (in press): Vitrinite reflectance as an indicator of hydrothermal metamorphism in liquid-dominated geothermal systems, Salton Trough, western North America, U.S. Geol. Survey Bulletin

BARKER, C.E. & W.A. ELDERS (1981): Vitrinite reflectance geothermometry and apparent heating duration in the Cerro Prieto geothermal field, Geothermics 10, 207-223

BARKER, C.E. & R.B. HALLEY (in press): Fluid inclusion, stable isotope, and vitrinite reflectance evidence for the thermal history of the Bone Spring Limestone, southern Guadalupe Mountains, Texas, in: D.L. GAUTIER (ed.), Relationship of Organic Matter and Mineral Diagenesis, Soc. of Economic Paleontologists and Mineralogists, Special Publication

BARTENSTEIN, H. & R. TEICHMÜLLER (1974): Inkohlungsuntersuchungen, ein Schlüssel zur Prospektierung von Paläozoischen Kohlenwasserstoff-Lagerstätten: Inkohlung und Erdöl, Fortschr. Geologie Rheinland u. Westfalen 24, 129-160

BEAUMONT, C., C.E. KEEN & R. BOUTILIER (1982): On the evolution of rifted continental margins: comparison of models and observations for the Nova Scotian margin, Geophysical Journal of the Royal Astron. Society 70, 667-715

BLANQUART, P. & E. MÉRIAUX (1975): Étude comparative du pouvoir reflecteur de veines, passées, lits, reinules, filets et grains de matière organique dispersée dans quelques sédiments du bassin houiller du Nord et du Pas-de-Calais, in: B. ALPERN (ed.), Pétrographie organique et potentiel pétrolier, 27-40

BLUMER, M. (1965): Organic pigments: their long term fate, Science 149, 722-726

BODMER, H.P. (1982): Beiträge zur Geothermie der Schweiz, PhD Thesis ETH Zürich, No. 7037, 201 p.

BOLDIZSÁR, T. (1973): Positive heat flow anomaly in the Carpathian Basin, Geothermics 2, 61-67

BOLDIZSÁR, T. (1978): Production of geothermal energy in Hungary, Part I, The geothermal anomaly of the Pannonian Basin (in Hungarian), Geonômia ês Bânyâszat 11, 233-254

BOSTICK, N.H. (1973): Time as a factor in thermal metamorphism of phytoclasts (coaly particles), Cong. Internat. Strat. Geol. Carbonif. 7th Krefeld (1971), Compte Rendu 2, 183-193

BOSTICK, N.H. (1974): Phytoclasts as indicators of thermal metamorphism, Franciscan assemblage and Great Valley sequence (Upper Mesozoic), California, Geol. Soc. America, Special Paper 153, 1-17

BOSTICK, N.H. (1979): Organic petrography of nineteen rocks, a split of each analyzed in thirty different laboratories (abst.), Abstracts of Papers, Ninth International Congress of Carboniferous Stratigraphy and Geology, Urbana, Illinois, May, 24

BOSTICK, N.H., S.M. CASHMAN, T.H. McCULLOH & C.T. WADELL (1979): Gradients of vitrinite reflectance and present temperature in the Los Angeles and Ventura Basins, California, in: D.F. OLTZ (ed.), Low temperature metamorphism of kerogen and clay minerals, 65-96, Los Angeles

BOSTICK, N.H. & J.N. FOSTER (1975): Comparison of vitrinite reflectance in coal seams and in kerogene of sandstones, shales and limestones in the same part of a sedimentary section, in: B. ALPERN (ed.), Pétrographie organique et potentiel pétrolier, 13-26

BOTTINGA, Y. & M. JAVOY (1973): Comments on isotope geothermometry, Earth Plan. Sci. Lett. 20, 250-265

BOWERS, T.S. & H.C. HELGESON (1983a): Calculation of the thermodynamic and geochemical consequences of nonideal mixing in the system H_2O-CO_2-NaCl on phase relations in geologic systems: equation of state for H_2O-CO_2-NaCl fluids at high pressures and temperatures, Geochim. Cosmochim. Acta 47, 1247-1275

BOWERS, T.S. & H.C. HELGESON (1983b): Calculation of the thermodynamic and geochemical consequences of nonideal mixing in the system H_2O-CO_2-NaCl on phase relations in geologic systems: metamorphic equilibria at high pressures and temperatures, Am. Min. 68, 1059-1075

BOWERS, T.S. & H.C. HELGESON (1985): Fortran programs for generating fluid inclusion isochores and fugacity coefficients for the system H_2O-CO_2-NaCl at high pressures and temperatures, Comp. and Geosci. 11, 203-213

BREITSCHMID, A. (1982): Diagenese und schwache Metamorphose in den sedimentären Abfolgen der Zentralschweizer Alpen (Vietwaldstätter See, Urirotstock), Ecl. Geol. Helv. 75, 331-380

BROOKS, J.D. (1970): The use of coals as indicators of the occurrence of oil and gas, Australian Petroleum Exploration Association Journal 10, 35-40

BROWNE, P.L. (1978): Hydrothermal alteration in active geothermal fields, Annual Reviews in Earth and Planetary Science 6, 229-250

BUCHLI, R. & D. WERNER (1985): Differential uplift history of the Central Alps, J. of Geoph., in revision

BUNTEBARTH, G. (1978): The degree of metamorphism of organic matter in sedimentary rocks as a paleo-geothermometer, applied to the Upper Rhine Graben, Pageoph. 117, 83-91

BUNTEBARTH, G. (1979): Eine empirische Methode zur Berechnung von paläogeothermischen Gradienten aus dem Inkohlungsgrad organischer Einlagerungen in Sedimentgesteinen mit Anwendung auf den mittleren Oberrhein-Graben, Fortschr. Geol. Rheinland u. Westfalen 29, 97-108

BUNTEBARTH, G. (1982): Geothermal history estimated from the coalification of organic matter, Tectonophysics 83 1-2, 101-108

BUNTEBARTH, G. (1983): Zur Paläogeothermie im Permokarbon der Saar-Nahe-Senke, Z. dt. geol. Ges. 134, 211-223

BUNTEBARTH, G. (1985): Das Temperaturgefälle im Dach des Bramscher Intrusivs bei Ibbenbüren aus Inkohlungsuntersuchungen, Fortschr. Geol. Rheinland u. Westfalen 33, 255-264

BUNTEBARTH, G., H. GREBE, M. TEICHMÜLLER & R. TEICHMÜLLER (1979): Inkohlungsuntersuchungen in der Forschungsbohrung Urach 3 und ihre geothermische Interpretation, Fortschr. Geol. Rheinland u. Westfalen 27, 183-199

BUNTEBARTH, G., I. KOPPE & M. TEICHMÜLLER (1982a): Palaeogeothermics in the Ruhr Basin, in: V. CERMAK & R. HAENEL (eds.), Geothermics and Geothermal Energy, 45-53

BUNTEBARTH, G., W. MICHEL & R. TEICHMÜLLER (1982b): Das permokarbonische Intrusiv von Krefeld und seine Einwirkung auf die Karbon-Kohlen am linken Niederrhein, Fortschr. Geol. Rheinland u. Westfalen 30, 31-45

BUNTEBARTH, G. & M. TEICHMÜLLER (1982): Ancient heat flow density estimated from the coalification of organic matter in the borehole Urach 3 (SW-Germany), in: R. HAENEL (ed.), The Urach Geothermal Project (Swabian Alb/Germany), 89-95, Stuttgart (Schweizerbart)

BURNE, R.V. & A.J. KANSTLER (1977): Geothermal constraints on the hydrocarbon potential of the Canning Basin, western Australia, J. Austral. Geol. Geophys. 2, 271-288

BURST, J.P. (1969): Diagenesis of Gulf Coast clayey sediments and its possible relation to petroleum migration, A.A.P.G. Bull. 53, 73-77

BYERS, J.D. & J.G. ERDMAN (1983): Low temperature degradation of carotenoids as a model for early diagenesis in recent sediments, in: M. BJORØY et al. (eds.), Advances in Organic Geochemistry (1981), 725-732 (Wiley)

CASTANO, J.R. & D.M. SPARKS (1974): Interpretation of vitrinite reflectance measurements in sedimentary rocks and determination of burial history using vitrinite reflectance and authigenic minerals, Geol. Soc. America, Special Paper 153, 31-52

CARRETA, Z. & M. WOLF (1980): The reflectance as rank parameter of brasilian Gondwana coals (in portuguese), Pesquisas 13, 35-42

CERMÁK, V. (1979): Heat flow map of Europe, in: V. CERMÁK & L. RYBACH (eds.), Terrestrial Heat Flow in Europe, 3-40, Berlin-Heidelberg-New York (Springer)

CERMÁK, V. & E. HURTIG (1979): Heat flow map of Europe, in: V. CERMÁK & L. RYBACH (eds.), Terrestrial heat flow in Europe, Intern. Union Comm. Geodyn., Sci. Rep., 58, Berlin-Heidelberg-New York (Springer)

CERMÁK, V. & J. ZAHRADNIK (1982): Two-dimensional correlation of heat flow and crustal thickness in Europe, in: V. CERMÁK & R. HAENEL (eds.), Geothermics and geothermal Energy, 17-25

CHAPMAN, R.E. (1973): Petroleum Geology, 304 p. (Elsevier)

CISSARZ, A. (1965): Einführung in die allgemeine und systematische Lagerstättenlehre, 228 p., Stuttgart (Schweizerbart)

CLARK, S.P.jr. & E. JAEGER (1969): Denudation rates in the Alps from geochronologic and heat flow data, A. Journ. Sc. 267, 1143-1160

CLAYTON, R.N. (1981): Isotopic thermometry, in: R.C. NEWTON, A. NAVROTSKY & B.J. WOOD (eds.), Thermodynamics of Minerals and Melts, 85, New York (Springer)

CLIMAP (1976): The surface of the ice-age earth, Science 191, 1131-1137

CLIMAP (1981): Seasonal reconstructions of the earth's surface at the last glacial maximum, Geological Society of America Map and Chart Series MC 36

C.N.R. (Consiglio Nazionale delle Richerche) (1980): Sezioni geologico-strutturali in scala 1:200.000 attraverso l'Appennino settentrionale, Progetto final. Geodinam., Sotto-Progetto 5, Mod. strutt. Gruppo App. sett., Firenze

CONNAN, J. (1974): Time-temperature relation in the oil genesis, A.A.P.G. Bull. 58, 2516-2521

CORNELIUS, C.D. (1975): Geothermal aspects of hydrocarbon exploration in the North Sea area, Norg. geol. unders. 316, 29-68

CUMMINGS, J.J. & W.E. ROBINSON (1972): Thermal degradation of Green River kerogen at 150° to 350° C, U.S. Bureau of Mines Rep. Invest. 7620

CURIALE, J.A., W.E. HARRISON & G. SMITH (1983): Sterane distribution of solid bitumen pyrolyzates. Changes with biodegradation of crude oil in the Ouachita Mountains, Oklahoma, Geochim. et Cosmochim. Acta 47, 517-523

CURRIE, J.B. & S.O. NWACHUKWU (1974): Evidence of incipient fracture porosity in reservoir rocks at depth, Bull. Can. Petr. Geol. 22, 42-58

DAHL, P.S. (1979): Comparative geothermometry based on major element and oxygen isotope distributions in Precambrian metamorphic rocks from southwestern Montana, Am. Mineral. 64, 1280-1293

DE BRAEMACKER, J.C. (1983): Temperature, subsidence, and hydrocarbon maturation in extensional basins: a finite element model, A.A.P.G. Bull. 67, 9, 1410-1414

DEINES, P. (1977): On the oxygen isotope distribution among mineral triplets in igneous and metamorphic rocks, Geochim. et Cosmochim. Acta 41, 1709-1730

DEMAISON, G.J. (1975): Relationship of coal rank to paleotemperatures in sedimentary rocks, in: B. ALPERN (ed.), Petrographie de la matiere organique des sediments, relations avec la paleotemperature et le potentiel petrolier, Paris, Centre Nat. de la Recherche Sci., 217-224

DENG, XIAO, WANG JI-AN (1982): Terrestrial heat flow in Anhui Province, in: Research on Geology (1), Institute of Geology, Academia Sinica, 82-89, Cultural Relics Publishing House, Beijing

DOW, W.G. (1977): Kerogen studies and geological interpretations, J. geochem. Explor. 7, 79-99

DÖVÉNYI, P., F. HORVÁTH, P. LIEBE, J. GÁLFI & I. ERKI (1983): Geothermal conditions of Hungary, Geophys. Transact., Budapest 29, 1, 1-58

DRURY, M.J. (1984): On a possible source of error in extracting equilibrium formation temperatures from borehole BHT data, Geothermics 13, 175-180

DRURY, M.J., A.M. JESSOP & T.J. LEWIS (1984): The detection of groundwater flow by precise temperature measurements in boreholes, Geothermics 13, 163-174

DUPLESSY, J.-C., L.L. LABEYRIE & N.J. SHACKLETON (1985): The oxygen isotope record of benthic foraminifera: effect of deep water temperature and ice volume changes, EOS 66, 291 (abstract)

DU ROUCHET, J. (1980): The program DIAGEN, two methods for considering the chemical evolution of organic matter, Bull. Centr. Rech. Expl.-Product., Elf-Aquit. 4, 813-831

EGGEN, S.S. (1984): Modelling of subsidence, hydrocarbon generation and heat transport in the Norwegian North Sea, IFP/Bernard Doronel, in press

ELDERS, W.A., D.K. BIRD, A.E. WILLIAMS & P. SCHIFFMAN (1984): Hydrothermal flow regime and magmatic heat source of the Cerro Prieto geothermal system, Baja California, Mexico, Geothermics 13, 1/2, 27-47

ELLIS, A.J., & W.A.J. MAHON (1977): Chemistry and Geothermal Systems, Acad. Press, New York, 392 p.

ENGLAND, P.C. (1978): Some thermal considerations of the Alpine-metamorphism - past, present, and future, Tectonophysics 46, 21-40

ENSMINGER, A. (1977): Evolution de composés polycycliques sedimentaires, Thesis, Doctorat-ès-Sciences, Univ. Louis Pasteur

ENSMINGER, A., G. JOLY & P. ALBRECHT (1978): Rearranged steranes in sediments and crude oils, Tetrahedron Lett. 18, 1575-1578

EPSTEIN, A.G., J.B. EPSTEIN & L.D. HARRIS (1977): Conodont color alteration - an index to organic metamorphism, U.S. Geological Survey Professional Paper 995, Washington (US Government Printing Office)

ESPITALIÉ, J. (1979): Charakterisierung der organischen Substanz und ihres Reifegrades in vier Bohrungen des mittleren Oberrheingrabens sowie Abschätzung der paläozoischen geothermischen Gradienten, Fortschr. Geol. Rheinland und Westfalen 27, 87-96

FALVEY, D.A. & L. DEIGHTON (1982): Recent advances in burial and thermal geohistory analysis, Journ. Austral. Petr. Expl. Assoc. 22, 65-81

FISCHER, A.G. (1969): Geological time-distance rates, the Bubnoff unit, Geol. Soc. America Bull. 80, 549-552

FLÜHMANN, P. (1976): Auswertung der Bohrlochmessungen (Sandsteinhorizonte und Kohlenflöze), Geol. Jb. A27, 21-28

FOURNIER, R.O. & J.J. ROWE (1966): Estimation of underground temperatures from the silica content of water from hot springs and wet steam wells, Am. Jour. Sci. 264, 658-697

FOURNIER, R.O. & A.H. TRUESDELL (1973): An empirical Na-K-Ca geothermometer for natural waters, Geochim. et Cosmochim. Acta 37, 1255-1275

FRECKMAN, J.T. (1978): Fluid inclusion and oxygen isotope geothermometry of rock samples from Sinclair # 4 and Elmore # 1 boreholes, Salton Sea geothermal field, Imperial Valley, California, USA, Master of Science Thesis, Univ. of California, Riverside, CA 92521, 66 p.

FREER, R. & P.F. DENNIS (1982): Oxygen diffusion studies, I. A preliminary ion microprobe investigation of oxygen diffusion in some rock forming minerals, Min. Mag. 45, 179-192

FREY, M., J.C. HUNZIKER, J.R. O'NEIL & H.W. SCHWANDER (1976): Equilibrium-disequilibrium relations in the Monte Rosa granite, Western Alps: Petrological, Rb-Sr and stable isotope data, Contr. Min. Petr. 55, 147-179

FREY, M., J.C. HUNZIKER, P. ROGGWILLER & C. SCHINDLER (1973): Progressive niedriggradige Metamorphose glaukonitführender Horizonte in den helvetischen Alpen der Ostschweiz, Contr. Min. Petr. 39, 185-218

FREY, M., M. TEICHMÜLLER, R. TEICHMÜLLER, J. MULLIS, B. KÜNZI, A. BREITSCHMIDT, U. GRUNER & B. SCHWIZER (1980): Very low-grade metamorphism in external parts of the Central Alps: illite crystallinity, coal rank and fluid inclusion data, Eclogae Geol. Helv. 73, 173-203

GALLEGOS, E.J. (1975): Terpane-sterane release from kerogen by pyrolysis gas chromatography-mass spectrometry, Anal. Chem. 47, 1524-1528

GATES, W.L. (1976): The numerical simulation of ice-age climate with a general circulation model, Journal of the Atmospheric Sciences 33, 1844-1873

GEHRIG, M. (1980): Phasengleichgewichte und PVT-Daten ternärer Mischungen aus Wasser, Kohlendioxid und Natriumchlorid bis 3 kbar und 550° C, Dissertation, Univ. Karlsruhe, Freiburg, 109 p. (Hochschulverlag)

Geothermal Research Division, Institute of Geology, Academia Sinica (1978): Regional Geothermal characteristics in the North China Plain, in: Collection of Works in Geothermal Studies, Beijing, 32-44 (Science Press)

Geothermal Research Division, Institute of Geology, Academia Sinica (1979a): Report on the data of terrestrial heat flow in the North China Plain and adjacent regions and its study, Scientia Geologica Sinica 1, 1-12

Geothermal Research Division, Institute of Geology, Academia Sinica (1979b): The first group of heat flow data measured in China, Acta Seismol. Sinica 1, 1, 91-107

Geothermal Research Division, Institute of Geology, Academia Sinica (1980): Geothermal characteristics in the North China Plain and its surroundings, in: Formation and development of the North China Fault Block Region, Beijing, 362-368 (Science Press)

GIESE, P. & A. STEIN (1971): Versuch einer einheitlichen Auswertung tiefenseismischer Messungen aus dem Bereich zwischen der Nordsee und den Alpen, Z. Geophysik 37, 237-272

GIESE, P., P. WIGGER, C. MORELLI & R. NICOLICH (1981): Seismische Studien zur Bestimmung der Krustenstruktur im Bereich der geothermischen Anomalie der Toskana, Schlußber. Forsch.progr. EC Contr. Nos. 486-78-1 EG D and 487-78-1 EG I, Freie Univ. Berlin, 108 p.

GIGGENBACH, W.F. & G.L. LYON (1977): The chemical and isotopic composition of water and gas from the Ngawha geothermal field, Open File Rep. 30/555/7 Chem. Div., Petone

GOLDMAN, D.S. & A.L. ALBEE (1977): Correlation of Mg/Fe partitioning between garnet and biotite with $^{18}O/^{16}O$ partitioning between quartz and magnetite, Am. J. Sci. 277, 750-767

VAN GRAAS, G., J.M.A. BAAS, B. VAN DE GRAAF & J.W. DE LEEUW (1982): Theoretical organic geochemistry, I. The thermodynamic stability of several cholestane isomers calculated by molecular mechanics, Geochim. et Cosmochim. Acta 46, 2399-2402

GRETENER, P.E. (1981): Geothermics, using temperature in hydrocarbon exploration, A.A.P.G. Education Course Note Series 17, 156

GRETENER, P.E. & C.D. CURTIS (1982): Role of temperature and time on organic metamorphism, A.A.P.G. Bull. 66, 8, 1124-1149

HABICHT, M. (1965/66): Die permo-karbonischen Aufschlußbohrungen der Nahe-Senke des Mainzer Beckens und der Zweibrücker Mulde, Z. dt. geol. Ges. 115, 631-649

HACQUEBARD, P.A. (1975): Pre- and post-deformational coalification and its significance for oil and gas exploration, in: Petrographie organique et potentiel petrolier, 225-241, Centre Nat. Res. Sci., Paris

HACQUEBARD, P.A. (1977): Rank of coal as an index of organic metamorphism of oil and gas in Alberta, Geol. Survey Bull. Canada 262, 11-23

HAENEL, R. & G. ZOTH (1982): Temperature measurements and determination of heat flow density, in: R. HAENEL (ed.), The Urach Geothermal Project (Swabian Alb/Germany), 81-88, Stuttgart (Schweizerbart)

HAMILTON, A.C. (1976): The significance of pattern on distribution shown by forest plants and animals in tropical Africa for the reconstruction of palaeoenvironments: a review, Palaeoecology of Africa 9, 63-97

HAMMEN, T. van der, G.C. MAARLEVELD, J.C. VOGEL & W.H. ZAGWIJN (1967): Stratigraphy, climatic succession and radiocarbon dating of the Last Glacial in the Netherlands, Geologie en Mijnbouw 46, 79-95

HAMMERSCHMIDT, K. & M. WAGENER (1983): K/Ar Bestimmungen an Biotiten aus den kristallinen Gesteinen der Forschungsbohrung Urach III, N. Jb. Min. Monatsh. 1983, 1, 35-48

HAMMERSCHMIDT, K., G.A. WAGNER & M. WAGENER (1984): Radiometric dating on research drill core Urach II: a contribution to its geothermal history, J. Geophys. 54, 97-105

HARWOOD, R.J. (1977): Oil and gas generation by laboratory pyrolysis of kerogen, Am. Assoc. of Petrol. Geologists Bull. 61, 2082-2102

HAYS, J.D., J. IMBRIE & N.J. SHACKLETON (1976): Variations in the earth's orbit: pacemaker of the ice ages, Science 194, 1121-1132

HEDEMANN, H.A. (1976): Die Gebirgstemperaturen in der Bohrung Saar 1 und ihre Beziehung zum geologischen Bau, Geol. Jb. A27, 433-454

HEDEMANN, H.A. & R. TEICHMÜLLER (1966): Stratigraphie und Diagenese des Oberkarbons in der Bohrung Münsterland 1, Z. dt. geol. Ges. 115, 787-825

HEEK, K.H. van, H. JÜNTGEN, K.F. LUFT & M. TEICHMÜLLER (1971): Aussagen zur Gasbildung in frühen Inkohlungsstadien auf Grund von Pyrolyseversuchen, Erdöl und Kohle 24, 566-572

HELING, D. & M. TEICHMÜLLER (1974): Die Grenze Montmorillonit/Mixed Layer-Minerale und ihre Beziehung zur Inkohlung in der Grauen Schichtenfolge des Oligozäns im Oberrheingraben, Fortschr. Geol. Reinland u. Westfalen 24, 113-128

HÉROUX, Y., A. CHAGNON & R. BERTRAND (1979): Compilation and correlation of major thermal maturation indicators, A.A.P.G. Bull. 63, 2128-2144

HO, T.T.Y. (1978): Manual of organic geochemical interpretation, Conoco, Ponca City, Oklahoma

HOEFS, J. (1983): Isotopic geothermometers, MS presented at IUGG Gen. Ass., Hamburg

HOEFS, J., G. MÜLLER & A.K. SCHUSTER (1982): Polymetamorphic relations in iron ores from the Iron Quadrangle, Brazil, the correlation of oxygen isotope variations with deformation history, Contr. Min. Petrol. 79, 241-251

HOERNES, S. & H. FRIEDRICHSEN (1978): Oxygen and hydrogen isotope study of the polymetamorphic area of the northern Ötztal-Stubai Alps (Tyrol), Contr. Min. Petrol. 72, 19-32

HOERNES, S. & E. HOFFER (1985): Stable isotope evidence for fluid-present and fluid-absent metamorphism in metapelites from the Damara Orogen, Namibia, Contr. Min. Petrol. 90, 322-330

HOFFMANN, C.F., A.S.MACKENZIE, C.A. LEWIS, J.R. MAXWELL, J.L. OUDIN, B. DURAND & M. VANDENBROUCKE (1984): A biological marker study of coals, shales and oils from the Mahakam delta, Kalimantan, Indonesia 1984, Chem. Geol. 42, 1-23

HOOD, A. C.C.M. GUTJAHR & R.L. HEACOCK (1975): Organic metamorphism and the generation of petroleum, A.A.P.G. Bull. 59, 986-996

HORVÁTH, F., P. DÖVÉNYI & P. LIEBE (1981): Geothermics of the Pannonian Basin, Earth Evolution Sciences 1, 285-291

HORVÁTH, F. & L. ROYDEN (1981): Mechanism for the formation of the Intra-Carpathian Basins: A review, Earth Evolution Sciences 1, 307-316

HOWER, J.C. & A. DAVIS (1981): Application of vitrinite reflectance anisotropy in the evaluation of coal metamorphism, Geological Society of America Bulletin 92, 350-366

HOWER, J., E.V. ESLINGER, M.E. HOWER & E.A. PERRY (1976): Mechanism of burial metamorphism of argillaceous sediments : 1 mineralogical and chemical evidence, GSA Bulletin 87, 725-737

HUCK, G. & J. KARWEIL (1955): Physikalisch-chemische Probleme der Inkohlung, Brennstoff-Chemie 1/2, 36, 1-11

HULSTON, J.R. (1976): Isotope work applied to geothermal systems at the Institute of Nuclear Sciences, New Zealand, Geothermics 5, 89-96

HUNT, J.M. (1979): Petroleum Geochemistry and Geology, San Francisco, 617 p. (Freeman)

HUNZIKER, J.C. (1974): Rb-Sr and K-Ar age determination and the Alpine tectonic history of the Western Alps, Mem. Ist. Geol. Min. Univ. Padova, Vol. XXXI

HUTTON, A.C., A.C. COOK, A.J. KANSTLER & D.M. McKIRDY (1980): Organic matter of oil shales, APEA Journal 20, 44-67

IIJIMA, A., K. AOYAGI & T. KAZAMA (1984): Diagenetic zeolite zone modified by recent high heat flow in Miti-Kuromatsunai hole, southwest Hokkaido, Japan (Geothermally modified zeolite zone), Proc. 6th Intern. Conf. on Zeolites, Reno, USA, 1983, 595-603 (Butterworth, Surrey)

IMBRIE, J. (1985): A theoretical framework for the Pleistocene ice ages, Journal of the Geological Society of London 142, 417-432

JACOB, H. & K. KUCKELKORN (1977): Das Inkohlungsprofil der Bohrung Miesbach 1 und seine erdölgeologische Interpretation, Erdöl-Erdgas Z. 93, 115-124

JACOB, H., K. KUCKELKORN & M. MÜLLER (1982): Inkohlung und Tektonik im Bereich der gefalteten Molasse, Erdöl u. Kohle 35, 510-518

JAEGER, J.C. (1964): Thermal effects of intrusions, Rev. Geophys. 2, 443-446

JAEGER, E., E. NIGGLI & E. WENK (1967): Altersbestimmungen an Glimmern der Zentralalpen, Beitr. Geol. Karte Schweiz, NF 134

JONES, P.H. (1970): Geothermal resources of the northern Gulf of Mexico basin, U.N. Symp. Pisa Compt. Rend.

JÜNTGEN, H. & J. KLEIN (1975): Formation of natural gas from coaly sediments, Erdöl u. Kohle 28, 65-73

JONES, J.M., D.G. MURCHISON & S.A. SALEH (1972): Variation of vitrinite reflectivity in relation to lithology, Adv. in Org. Geoch. 1971, 601-612

KAMILLI, R.J. & H. OHMOTO (1977): Paragenesis, zoning, fluid inclusion and isotopic studies of the Finlandia Vein, Colqui District, Central Peru, Econ. Geol. 72, 950-982

KANTSLER, A.J. G.C. SMITH & A.C. COOK (1978): Lateral and vertical rank variation: implications for hydrocarbon exploration, Austral. Petrol. Explor. Assoc. J. 18, 143-156

KAPPELMEYER, O. & R. HAENEL (1974): Geothermics, with special reference to application, Geoexpl. Monograph Ser. 1, No. 4, Berlin, 1-238 (Borntraeger)

KARPOV, P.A., A.F. STEPANOVA, N.V. SOLOVEVA, A.P. AGULOV, A.L. GOZHAJA & V.P. TSAITSKY (1975): Quantitative analysis of the temperature and geological time as factors of the coalification of dispersed coaly particles and its possible use in the petroleum geology (in Russian), Izv. AN URSS, Ser. Geol. 3, 103-113

KARWEIL, J. (1956): Die Metamorphose der Kohlen vom Standpunkt der physikalischen Chemie, Z. Deutscher Geol. Ges. 107, 132-139

KARWEIL, J. (1975): The determination of paleotemperatures from the optical reflectance of coaly particles, in: B. ALPERN (ed.), Pétrographie de la matière organique des sédiments, CNRS, Paris, 195-203

KATZ, B.J., L.M. LIRO, J.E. LACEY & H.W. WHITE (1982): Time and temperature in petroleum formation: application of Lopatin's method to petroleum exploration, Discussion Amer. Ass. Petr. Geol. Bull. 66, 8, 1150-1152

KATZ, H.R. (1979): Alpine uplift and subsidence of foredeeps, in the origin of the southern Alps, Royal Soc. New Zealand Bull. 18, 121-130

KELLEY, S., I. DUNCAN & D. BLACKWELL (1983): Use of fission-track annealing systematics in constraining the thermal evolution of sedimentary basins (an abstract), A.A.P.G. Bull. 67, 494

KETTEL, D. (1981): Maturitätsberechnung für das nordwestdeutsche Oberkarbon - ein Test verschiedener Methoden, Erdöl-Erdgas Z. 97, 11, 395-404

KHARAKA, Y.K. et al. (1980): Geochemistry of formation waters from Pleasant Bayou No 2, Well, Proc. 4th Geopressure-Geothermal Energy Conf. Austin, Univ. Texas, 168-193

KISCH, H.J. (1969): Coal-rank and burial metamorphic mineral facies, in: P.A. SCHENCK & I. HAVENNAAR (eds.), Advances in Organic Geochemistry (1968), Oxford, 407-425 (Pergamon Press)

KISSLING, E., T.P. LABHART & L. RYBACH (1978): Radiometrische Untersuchungen an Rotondogranit, Schweiz. mineral. petrogr. Mitt. 58, 357-388

KOPPE, I. (1980): Palaeogeothermische Untersuchungen im Ruhrgebiet und am Niederrhein anhand von Inkohlungsgraden organischer Substanzen, Dipl. Arbeit, Techn. Univ. Clausthal, 73 p.

KREULEN, R. & P.C.J.M.van BEEK (1983): The calcite-graphite isotope thermometer, data on graphite-bearing marbles from Naxos, Greece, Geochim. et Cosmochim. Acta 47, 1527-1530

KUCKELKORN, K. & H. JACOB (1977): Inkohlungsgradienten im Alpenvorland, Compendium 77/78, Erdöl u. Kohle, 331-338

KÜBLER, B. (1967): La cristallinité de l'illite et les zones tout à fait supérieures du metamorphisme, Etages tectoniques, Colloq. Neuchâtel, 105-122

KÜNSTNER, E. (1974): Vergleichende Inkohlungsuntersuchungen unter besonderer Berücksichtigung mikrophotometrischer Reflexionsmessungen an Kohlen, Brandschiefern und kohlehaltigen Nebengestein, Freib. Forschungsh. C287

LAMB, H.H. (1977): Climate Present, Past and Future, Vol. 2: Climatic History and the Future, London, 835 p. (Methuen)

LANDAU, L.D. & E.M. LIFSCHITZ (1978): Hydrodynamik, Lehrbuch der theoretischen Physik, Bd. 6, Berlin (Akademie-Verlag)

LANDIS, C.A. (1971): Graphitization of dispersed carbonaceous material in metamorphic rocks, Contr. Min. Petrol. 30, 34-45

LASAGA, A. (1981): Rate laws of chemical reactions, in: Kinetics of geochemical processes, Reviews in Mineralogy 8, 1-68

LEFLER, J. & Cs. SAJGÓ (1985): Limits of application of the reaction kinetic method in paleogeothermics (in print)

LEMMLEIN, G.G. (1956): Formation of fluid inclusions in minerals and their use in geological thermometry, Geochemistry 6, 630-642

LERCHE, I., R.F. YARZAB & C.G.ST.C. KENDALL (1984): Determination of paleoheat flux from vitrinite reflectance data, A.A.P.G. Bull. 68, 1704-1717

LEROY, J. (1979): Contribution à l'etalonage de la pression interne des inclusions fluides lors de leur decrepitation, Bull. Mineral. 102, 584-593

LINDSTRÖM, M. (1964): Conodonts, Amsterdam-London-New York (Elsevier)

LOPATIN, N.V. (1971): Temperature and geological time as factors of carbonification, Akad. Nauk SSSR Izv. Ser. Geol. 3, 95-106

LOPATIN, N.V. (1976): Influence of temperature and geological time on the catagenetic processes of coalification and petroleum generation in study of organic matter in recent and ancient sediments (in Russian), N.B. VASSOJEVICH (ed.), 361-366

LOPATIN, N.V. (1976): Istoriko-geneticheskiy analiz nefteobrazovaniya s ispolzovaniem modeli ravnomernovo neprerivnovo opuskaniya neftematerinskovo plasta (Historico-genetic analysis of petroleum generation: Application of a model of uniform continuous subsidence of the oil-source bed), AN SSSR Izv. Ser. Geol. $\underline{8}$, 93-101

LOPATIN, N.V. & N.H. BOSTICK (1973): Geologicheskiye faktory katagneza ugley, Akad. Nauk SSSR Otdeleniye Geologii, Geofiziki, Geokhimi, Kom. Osad. Porodam, Moskow, 79-90 (Nauka Press), English Translation Illinois Geological Survey Reprint 1974-Q, 16 p.

LUDWIG, B., G. HUSSLER, P. WEHRUNG & P. ALBRECHT (1979): C_{26}-C_{29} triaromatic steroid derivatives in sediments and petroleums, Tetrahedron Lett., 3313-3316

LUDWIG, B., G. HUSSLER, P. WEHRUNG & P. ALBRECHT (1981): Identification of mono- and triaromatic steroid derivatives in ancient shales and petroleum, Tenth Internat. Meeting on Organic Geochemistry, Abstracts: M. BJORØY (ed.), 59

MACKENZIE, A.S. (1980): Applications of biological marker compounds to subsurface geological processes, Ph.D. Thesis, Univ. of Bristol

MACKENZIE, A.S., S.C. BRASSEL, G. EGLINTON & J.R. MAXWELL (1982): Chemical fossils, the geological fate of steroids, Science $\underline{217}$, 491-504

MACKENZIE, A.S., C.F. HOFFMANN & J.R. MAXWELL (1981a): Molecular parameters of maturation in the Toarcian shales, Paris Basin, France-III, Changes in aromatic steroid hydrocarbons, Geochim. et Cosmochim. Acta $\underline{45}$, 1345-1355

MACKENZIE, A.S., C.A. LEWIS & J.R. MAXWELL (1981b): Molecular parameters of maturation in the Toarcian shales, Paris Basin, France-IV, Laboratory thermal alteration studies, Geochim. et Cosmochim. Acta $\underline{45}$, 2369-2376

MACKENZIE, A.S. & D. McKENZIE (1983): Isomerization and aromatization of hydrocarbons in sedimentary basins formed by extension, Geological Magazine $\underline{120}$, 417-528

MACKENZIE, A.S. R.L. PATIENCE, J.R. MAXWELL, M. VANDENBROUCKE & B. DURAND (1980): Molecular parameters of maturation in the Toarcian shales, Paris Basin-I. Changes in the configurations of acyclic. isoprenoid alkanes, steranes and triterpanes, Geochim. et Cosmochim. Acta $\underline{44}$, 1709-1721

MACKENZIE, A.S. L. REN-WEI, J.R. MAXWELL, J.M. MOLDOWAN & W.K. SEIFERT (1983): Molecular measurements of thermal maturation of cretaceous shales from the Overthrust Belt, Wyoming, USA, in: M. BJORØY et al. (eds.), Advances in Organic Geochemistry (1981), 496-503 (Wiley)

MAGARA, K. (1978): Compaction and Fluid Migration, Practical Petroleum Geology, Developments in Petroleum Science $\underline{9}$, 319 p.

MAHON, W.A.J. (1966): Silica in hot water discharged from drillholes at Wairakei, New Zealand, N.Z. Jour. Sci. $\underline{9}$, 125-144

MAJOROWICZ, J.A. & A.M. JESSOP (1981): Present heat flow and a preliminary paleogeothermal history of the Central Prairies basin, Canada, Geothermics $\underline{10}$, 2, 81-93

Map of Oil and Gas Prospection in Hungary (1978): Publication of the National Oil and Gas Trust in coop. with the Hung. Geol. Survey, Budapest

MATSUHISA, Y., J.R. GOLDSMITH & R.N. CLAYTON (1978): Mechanism of hydrothermal crystallization of quartz at 250° C and 15 kbar, Geochim. et Cosmochim. Acta 42, 173-182

MATSUHISA, Y., J.R. GOLDSMITH & R.N. CLAYTON (1979): Oxygen isotope fractionation in the systems quartz-albite-anorthite-water, Geochim. et Cosmochim. Acta 43, 1131-1140

MATTHEWS, A., J.R. GOLDSMITH & R.N. CLAYTON (1983a): Oxygen isotope fractionation involving pyroxenes: the calibration of mineral-pair geothermometers, Geochim. et Cosmochim. Acta 47, 631-644

MATTHEWS, A., J.R. GOLDSMITH & R.N. CLAYTON (1983b): Oxygen isotope fractionation between zoisite and water, Geochim. et Cosmochim. Acta 47, 645-654

McCARTNEY, J.T. & M. TEICHMÜLLER (1972): Classification of coals according to degree of coalification by reflectance of the vitrinite component, Fuel 51, 64-68

McKENZIE, D.P. (1978): Some remarks on the development of sedimentary basins, Earth Planet. Sci. Lett. 40, 25-32

McKENZIE, D.P. (1981): The variation of temperature with time and hydrocarbon maturation in sedimentary basins formed by extension, Earth Planet. Sci. Lett. 55, 87-98

McKENZIE, D., A.S. MACKENZIE, J.R. MAXWELL & Cs. SAJGÓ (1983): Isomerization and aromatization of hydrocarbons in stretched sedimentary basins, Nature 301, 504-506

McNAB, J.G., P.V. SMITH & R.L. BETTS (1952): The evolution of petroleum, Industrial and Engineering Chemistry 44, 2556-2563

McTAVISH, R.A. (1978): Pressure retardation of vitrinite reflectance, offshore northwest Europe, Nature 271, 648-650

MIDDLETON, M.F. (1982): Tectonic history from vitrinite reflectance, Geophys. J.R. astr. Soc. 68, 121-132

MIDDLETON, M.F. & D.A. FALVEY (1983): Maturation modeling in Otway basin, Australia, A.A.P.G. Bull. 67, 2, 271-279

MIDDLETON, M.F. & T.G. RUSSELL (1981): Coal rank and organic diagenesis studies in the Gunnedah Basin: a preliminary report, Quart. Notes Geol. Surv. N.S.W., October (Complete Issue)

MILANKOVITCH, M. (1941,1969): Kanon der Erdbestrahlung und seine Anwendung auf das Eiszeitenproblem, Royal Serbian Academy Spec. Publ. 133; English translation Israel Program for Scientific Translations

MILLER, G.H., H.P. SEJRUP, J. MANGERUD & B.G. ANDERSEN (1983): Amino acid ratios in Quaternary molluscs and foraminifera from Western Norway: correlation, geochronology and paleotemperature estimates, Boreas 12, 107-124

MILLIKEN, K.L., L.S. LAND & R.G. LOUCKS (1981): History of burial diagenesis determined from isotopic geochemistry, Frio formation, Brazoria county, Texas, A.A.P.G. Bull. 65, 1397-1413

MITSUI, K. & K. TAGUCHI (1977): Silica mineral diagenesis in Neogene Tertiary shales in the Tempoku district, Hokkaido, Japan, J. Sed. Petrol. 47, 158-167

MOSCVIN, V.I. (1983): About some phenomena accompanying the formation of oil in Bazhenovskaya Formation of Western Siberia (in Russian), Geologiya i Geofizika No 11 (287), 54-61

MUCSI, M. & J. RÉVÉSZ (1975): Neogene evolution of the Southeastern part of the Great Hungarian Plain on the basis of sedimentological investigations, Acta Miner. Petr. XXII, 1, 29-49

MULHEIRN, L.S. & G. RYBACK (1977): Isolation and structure analysis of steranes from geological sources, in: R. CAMPOS & J. GARI (eds.), Advances in Organic Geochemistry (1975), Madrid, 173-192 (ENADISMA)

MUNDRY, E. (1968): Über die Abkühlung magmatischer Körper, Geol. Jb. Hannover 95, 755-766

MÜLLER, G., A.K. SCHUSTER & J. HOEFS (1982): Oxygen isotope variations in polymetamorphic iron ores from the Quadrilatero Ferrifero, Brazil, Revista Brazil. Geosie. 12, 348-355

MÜLLER, M. (1975): Ergebnisse der ersten, im Rahmen des Erdgasaufschlußprogramms der Bundesregierung mit öffentlichen Mitteln geförderten Erdgastiefbohrung, Erdöl u. Kohle, 63-76

NAESER, N.D. (1984): Fission-track ages from the Wagon Wheel No. 1 well, Northern Green River Basin, Wyoming, Evidence for recent cooling, U.S. Geological Survey Open-file Report 84-753, 66-77

NAGORNYI, V.N. & Yu. N. NAGORNYI (1974): The question of the quantitative evaluation of the role of the time in processes of the regional metamorphism of coals, Solid Fuel Chemistry 8, 30-36

NERUCHEV, S.G. & G.M. PARPAROVA (1972): The role of geologic time in processes of metamorphism of coal and dispersed organic matter in rocks (in Russian), Otdeleniye Geologiya i Geofizika, Akad. Nauk.SSSR Sibirsk 10, 3-10

NERUCHEV, S.G., E.A. ROGOZINA, I.A. ZELICHENKO & P.A. TRUSHKOV (1980): Geokhimicheskije osobennostni processov nefte i gazoobrazovaniya v otlozhemiyak bazhenovskoj sviti Zapadno-Sibirskoy nizmennosti Izv., AN. SSSR, Ser. Geol. 2, 5-16

NETO, C.C. (1983): Theoretical organic geochemistry I. An alternative model for the epimerization of hydrocarbon chiral centers in sediments, in: M. BJORØY et al. (eds.), Advances in Organic Geochemistry (1981), 834-838

NEWMAN, J. & N.A. NEWMAN (1982): Reflectance anomalies in Pike River coals: evidence of variability in vitrinite type, with implications for maturation studies and "Suggate rank", New Zealand of Geol. and Geoph. 25, 2, 233-243

NIKOLOV, Z. (1974): Coalification in the stratigraphic profile of the Carboniferous of the Dobrudja Basin (in Bulgarian), Bull. Geol. Int. Ser. Petrol. Coal Geol. XXII, 135-152

NORTON, D. & J. KNIGHT (1977): Transport phenomena in hydrothermal systems: cooling plutons, Amer. J. Sci. 277, 937-981

OHMOTO, H. & R.O. RYE (1970): The Bluebell Mine, British Columbia I. Mineralogy, paragenesis, fluid inclusions and the isotopes of hydrogen, oxygen and carbon, Econ. Geol. 65, 417-437

OHMOTO, H. & R.O. RYE (1979): Isotopes of sulfur and carbon. Chapter 10 in: H.L. BARNES (ed.), Geochemistry of Hydrothermal Ore Deposits, New York (Wiley)

OTTENJANN, K., M. TEICHMÜLLER & M. WOLF (1974): Spektrale Fluoreszenz-Messungen an Sporiniten mit Auflicht-Anregung, eine mikroskopische Methode zur Bestimmung des Inkohlungsgrades gering inkohlter Kohlen, Fortschr. Geol. Rheinland u. Westfalen 24, 1-36

OURISSON, G., P. ALBRECHT & M. ROHMER (1979): The hopanoids. Palaeochemistry and biochemistry of a group of natural products, Pure Appl. Chem. 51, 709-729

OXBURGH, E.R. & D.L. TURCOTTE (1974): Thermal gradients and regional metamorphism in overthrust terrains with special reference to the Eastern Alps, Schweiz. mineral. petrogr. Mitt. 54

PACES, T. (1975): A systematic deviation from Na-K-Ca geothermometer below 75° C and above 10^{-4} atm. P_{CO_2}, Geochim. et Cosmochim. Acta 39, 541-544

PAGEL, M. (1975): Determination des conditions physico-chimiques de la silification diagenetique des gres Athabasca (Canada) au moyen des inclusions fluides, C.R. Acad. Sc. Paris 280 D, 2301-2304

PAGEL, M. & B. POTY (1983): The evolution of composition, temperature and pressure of sedimentary fluids over time: a fluid inclusion reconstruction, Colloque international phenomenes thermiques dans les bassins sedimentaires, 6-10 Juin 1983, ADERA, Saint-Médard-en-Jalles/France, Paris (Editions Technip, in press)

PANZA, G.F. & St. MUELLER (1978): The plate boundary between Eurasia and Africa in the Alpine area, Mem. Ist. Geol. Min. Univ. Padova XXXIII, 43-50

PATTEISKY, K., M. TEICHMÜLLER & R. TEICHMÜLLER (1962): Das Inkohlungsbild des Steinkohlengebirges an Rhein und Ruhr, dargestellt im Niveau von Flöz Sonnenschein, Fortschr. Geol. Rheinland u. Westf. 3, 2, 687-700

PERRY, E.A. & J. HOWER (1970): Burial diagenesis in Gulf Coast pelitic sediments, Clays and clay minerals 18, 165-177

PERRY, E.A. & J. HOWER (1972): Late-stage dehydration in deeply buried pelitic sediments, Am. Ass. Petrol. Geol. Bull. 56, 2013-2021

PETERS, K.E., R. ISHIWATARI & I.R. KAPLAN (1977): Color of kerogen as index of organic maturity, A.A.P.G. Bull. 61, 4, 504-510

PETROV, Al.A., S.D. PUSTIL'NIKOVA, N.N. ABRYUTINA & G.R. KAGRAMANOVA (1976): Neftjaniye szterani i triterpani (Petroleum steranes and triterpanes), Neftekhimiya 16, 411-427

PHILIPPI, G.T. (1965): On the depth, time and mechanism of petroleum generation, Geochim. et Cosmochim. Acta 29, 1021-1044

PIERI, M. & G. GROPPI (1981): Subsurface geological structure of the Po Plain, Italy, C.N.R. Progretto final. Geodinam., Mod. strutt. 414, 13 p.

PIETZNER, H., J. VAHL, H. WERNER & W. ZIEGLER (1968): Zur chemischen Zusammensetzung und Mikromorphologie der Conodonten, Palaeontographica 128 A, 115-152

POLLASTRO, R.M. & C.E. BARKER (in press): Comparative measures of paleotemperature - an example from clay-mineral, vitrinite reflectance, and fluid inclusion studies, Pinedale anticline, Green River basin, Wyoming, in: D.L. GAUTIER (ed.), Relationship of Organic Matter and Mineral Diagenesis, Society of Economic Paleontologists and Mineralogists, Special Publication

POTTER, R.W., II (1977): Pressure corrections for fluid inclusion homogenization temperatures based on the volumetric properties of the system $NaCl-H_2O$, J. Research, U.S. Geological Survey 5, 603-607

POTY, B., J. LEROUX & L. JACHIMOWICZ (1976): Un nouvel appareil pour la mesure des temperatures sous le microscope, Bull. Soc. Francais Mineral. Cristallogr. 99, 182-186

POTY, B. & M. PAGEL (1983): Fluid inclusion techniques to the study of thermal evolution of sedimentary basins, Colloque international phenomenes thermiques dans les bassins sedimentaires, 6-10 Juin 1983, ADERA, Saint-Médard-en-Jalles/France, Paris (Éditions Technip, in press)

POWERS, M.C. (1967): Fluid-release mechanisms in compacting marine mudrocks and their importance in oil exploration, Am. Ass. Petrol. Geol. Bull. 51, 1240-1254

PRICE, L.C. (1982): Time as a factor in organic metamorphism and use of vitrinite reflectance as an absolute paleogeothermometer, A.A.P.G. Bull. 66, 619-620

PRICE, L.C. (1983): Geologic time as a parameter in organic metamorphism and vitrinite reflectance as an absolute paleogeothermometer, Journ. of Petroleum Geology 6, 5-38

PRICE, L.C. & C.E. BARKER (1985): Suppression of vitrinite in amorphous-rich kerogen - a major unrecognized problem, Journal of Petroleum Geology 8, 59-84

PRICE, L.C., J.L. CLAYTON & L.L. RUMEN (1981): Organic geochemistry of the 9.6 km Betha Rodgers no. 1 well, Oklahoma, Organic Geochemistry 3, 59-77

PRODEHL, C., J. ANSORGE, J.B. EDEL, D. EMTER, K. FUCHS, St. MÜLLER & E. PETERSCHMIDT (1976): Explosion-seismology research in the central and southern Rhine Graben - a case history, in: P. GIESE, C. PRODEHL & A. STEIN (eds.), Explosion Seismology in Central Europe, Berlin-Heidelberg-New York, 313-328 (Springer)

PURDY, J.W. & E. JÄGER (1976): K-Ar ages on rock-forming minerals from the Central Alps, Mem. 1st Geol. Miner. dell Universita Padova 30, 1-32

PUSEY, C.W. (1973): How to evaluate potential gas and oil source rocks, World Oil, April, 71-76

PUSTIL'NIKOVA, S.D., N.N. ABRYUTINA, G.R. KAGRAMANOVA & Al.A. PETROV (1976): Hydrocarbons of the hopane series in oils, Geokhimya 1976, 460-468

REDLICH, O. & J.N.S. KWONG (1949): An equation of state, Fugacity of gaseous solutions, Chem. Rev. 44, 233-244

REUTTER, K.-J., M. TEICHMÜLLER, R. TEICHMÜLLER & G. ZANZUCCHI (1978): Coalification studies in the northern Apennines and palaeogeothermal implications, in: H. CLOSS, D. ROEDER & K. SCHMIDT (eds.), Alps, Apennines, Hellenides, Inter-Union Comm. Geodyn. Scient. Rep. 38, 261-267, Stuttgart (Schweizerbart)

REUTTER, K.-J., M. TEICHMÜLLER, R. TEICHMÜLLER & G. ZANZUCCHI (1983): The coalification pattern in the northern Apennines and its palaeogeothermic and tectonic significance, Geol. Rdsch. 72, 861-893

ROBERT, P. (1985): Histoire géothermique et diagenèse organique, Bull. Centres Rech. Explor. Prod. Elf-Aquitaine, Mém. 8, Pau

ROBERTS, W.H. (1980): Design and function of oil and gas traps, Problems of Petroleum Migration, A.A.P.G. Studies in Geology 10, 217-240

ROEDDER, E. (1962): Studies of fluid inclusions I., Econ. Geol. 57, 1045-1061

ROEDDER, E. (1963): Studies of fluid inclusions II., Econ. Geol. 58, 167-211

ROEDDER, E. (1967): Fluid inclusions as samples of ore fluids, in: H.L. BARNES (ed.), Geochemistry of hydrothermal ore deposits, New York, 515-574 (Holt, Rinsehart & Winston)

ROEDDER, E. (1976): Fluid inclusion evidence on the genesis of ores in sedimentary and volcanic rock, in: K.H. WOLF (ed.), Handbook of Stratabound and Stratiform Ore Deposits 2, 67-110, Amsterdam, New York (Elsevier)

ROEDDER, E. (1982): Igneous fluid inclusion geothermometry, EOS: Trans. Am. Geoph. Union 63, 471

RONSARD, J.N. & A. OBERLIN (1984): Contribution of high resolution electron microscopy to organic materials characterization and interpretation of their reflectance, MS

ROYDEN, L., F. HORVÁTH & J. RUMPLER (1983): Evolution of the Pannonian basin system: Part I: Tectonics, Tectonics 2 (1), 63-90

ROYDEN, L., & C.E. KEEN (1980): Rifting process and thermal evolution of the continental margin of eastern Canada determined from subsidence curves, Earth Planet. Sci. Lett. 51, 343-361

ROYDEN, L., J.G. SCLATER & R.P. von HERZEN (1980): Continental margin subsidence and heat flow: Important parameters in formation of petroleum hydrocarbons, A.A.P.G. Bull. 64, 2, 173-187

RUBINSTEIN, I., C. SPYCKERELLE & O.P. STRAUSZ (1979): Pyrolysis of asphaltenes - a source of geochemical information, Geochim. et Cosmochim. Acta 43, 1-6

RULLKÖTTER, J., Z. AIZENSHTAT & B. SPIRO (1984): Biological markers in bitumens and pyrolyzates of Upper Cretaceous bituminous chalks from the Chareb Formation (Israel), Geochim. et Cosmochim. Acta 48, 151-157

RULLKÖTTER, J. & P. PHILP (1981): Extended hopanes up to C_{40} in Thornton bitumen, Nature 292, 616-618

RUMBLE, D. (1982): Stable isotope fractionation during metamorphic devolatilization reactions, in: J.M. FERRY (ed.), Characterization of metamorphism through mineral equilibria, Reviews in Mineralogy 10, 327-353

RYBACH, L. (1984): The paleogeothermal conditions of the Swiss Molasse Basin, Inst. Francais Petr. Rev. 39, 2, 143-146

RYE, R.O. (1974): Comparison of sphalerite-galena sulfur isotope temperatures with filling-temperatures of fluid inclusions, Econ. Geol. 69, 26-32

SAJGÓ, Cs. (1980): Hydrocarbon generation in a super-thick Neogene sequence in Southeast Hungary. A study of the extractable organic matter, in: A.G. DOUGLAS & J.R. MAXWELL (eds.), Advances Geochemistry 1979, 103-113 (Pergamon Press)

SAJGÓ, Cs. (1984): Organic geochemistry of crude oils from South-east Hungary, in: P.A. SCHENCK, J.W. de LEEUW & G.W.M. LIJMBACH (eds.), Advances in Organic Geochemistry (as Org. Geochem. Vol. 6), 569-578 (Pergamon Press)

SAJGÓ, Cs., Z.A. HORVÁTH & J. LEFLER (1983): An organic maturation study of the Hód-I. borehole (Pannonian Basin), submitted to an AAPG Memoir Volume on Pannonian Basin

SAJGÓ, Cs. & J. LEFLER (1985): A reaction kinetic approach to the temperature-time history of sedimentary basins (in this volume)

SAJGÓ, Cs., J. LEFLER & Z.A. HORVÁTH (1983): Applicability of different organic maturation parameters, MS presented at IUGG XVIII. Gen. Ass. Hamburg

SAJGÓ, Cs., A.S. MACKENZIE & J.R. MAXWELL (1985): Changes in biological marker distribution in a thick Neogene sequence in Hungary, Org. Geochem. (accepted)

SAJGÓ, Cs., J.R. MAXWELL & A.S. MACKENZIE (1983): Evaluation of fractionation effects during the early stages of primary migration, Org. Geochem. $\underline{5}$, 65-73

SANFORD, S.J. (1981): Dating thermal events by fission track annealing, Cerro Prieto geothermal field, Baja California, Mexico, Institute of Geophysics and Planetary Physics, University of California, Riverside, UCR/IGPP Report

SANFORD, S.J. & W.A. ELDERS (1981): Dating thermal events at Cerro Prieto using fission track annealing, Proc. Third Symp. Cerro Prieto Geothermal Field, San Francisco, Lawrence Berkeley Lab.

SAVIN, S.M. (1977): The history of the earth's surface temperature during the past 100 million years, Annual Review of Earth Planetary Sciences $\underline{5}$, 319-355

SAVIN, S.M. & M. LEE (1984): Estimation of subsurface temperature from oxygen isotope ratios of minerals, MS.

SCHAEFLÉ, J. (1979): Thèse de Doctorat-ès-Sciences, Univ. Louis Pasteur

SCHMIDT-THOMÉ, P. (1957): Molasse-Untergrund und Helvetikum-Nordgrenze im Tegernsee-Bereich und die Herkunft von Erdöl und Jodwasser in Oberbayern, Geol. Jb. $\underline{74}$, 225-242

SCHWAB, F.L. (1976): Modern and ancient sedimentary basins, comparative accumulation rates, Geology $\underline{4}$, 723-727

SCLATER, J.G., L. ROYDEN, F. HORVÁTH, B.C. BURCHFIEL, S. SEMKEN & L. STEGENA (1980): The formation of the Intra-Carpathian basins as determined from subsidence data, Earth Planet. Sci. Lett. $\underline{51}$, 139-162

SEIFERT, W.K. (1978): Steranes and terpanes in kerogen pyrolysis for correlation of oils and source rocks, Geochim. et Cosmochim. Acta $\underline{42}$, 473-484

SEIFERT, W.K., R.M.K. CARLSON & J.M. MOLDOWAN (1983): Geomimetic synthesis, structure assignment, and geochemical correlation application of monoaromatized petroleum steroids, in: M. BJORØY et al. (eds.), Advances in Organic Geochemistry (1981), 710-724 (Wiley)

SEIFERT, W.K., & J.M. MOLDOWAN (1979): The effect of biodegradation on steranes and terpanes in crude oils, Geochim. et Cosmochim. Acta $\underline{43}$, 111-126

SEIFERT, W.K. & J.M. MOLDOWAN (1980): The effect of thermal stress on source rock quality as measured by hopane stereochemistry, in: A.G. DOUGLAS & J.R. MAXWELL (eds.), Advances in Organic Geochemistry (1979), 229-237 (Pergamon Press)

SEYER, W.F. (1933): The conversion of fatty and waxy substances into petroleum hydrocarbons, Institute of Petroleum Technologists Journal $\underline{9}$, 773-783

SHACKLETON, N.J. (1985): Oxygen isotope evidence for Cenozoic climatic change, in: P. BRENCHLEY (ed.), Fossils and Climate

SHACKLETON, N.J. & J.M. CHAPPELL (1985): The ocean deepwater oxygen isotope record and the New Guinea sea-level record, EOS 66 293 (abstract)

SHIBAOKA, M. & A.J.R. BENNETT (1977): Patterns of diagenesis in some Australian sedimentary basins, Austral. Petrol. Explor. Assoc. J. 17, 58-63

SHI JIYANG, A.S. MACKENZIE, R. ALEXANDER, G. EGLINTON, A.P. GOWAR, G.A. WOLF & J.R. MAXWELL (1982): A biological marker investigation of petroleums and shales from the Shengli oilfield, the People's Republic of China, Chemical Geology 35, 1-31

SHIMOYAMA, T. & A. IIJIMA (1976): Influence of temperature on coalification of Tertiary coal in Japan - summary, Circum-Pacific Energy and Mineral Resources, Memoir 25, American Association of Petroleum Geologists, 98-103

SIEVER, A. (1983): Burial history and diagenetic reaction kinetics, A.A.P.G. Bull. 67, 4, 684-691

SIGURDSON, D.R. (1974): Mineral paragenesis and fluid inclusion thermometry at four western U.S. tungsten deposits, Ph.D. thesis, University of California, Riverside, 214 p.

SIMMONS, G. (1967): Interpretation of heat flow anomalies. 2. Flux due to initial temperature of intrusives, Rev. Geophys. 5, 2, 109-120

SNEDECOR, G.W. & W.G. COCHRAN (1967): Statistical Methods, Ames. 593 p. (Iowa State University Press)

SNOWDON, L.R. (1979): Errors in extrapolation of experimental kinetic parameters to organic geochemical systems, A.A.P.G. Bull. 63, 1128-1134

STACH, E., M.-Th. MACKOWSKY, M. TEICHMÜLLER, G.H. TAYLOR, D. CHANDRA & R. TEICHMÜLLER (1982): Stach's Textbook of Coal Petrology, Third edition, Berlin, 535 p. (Gebrüder Borntraeger)

STALDER, P.J. (1979): Organic and inorganic metamorphism in the Taveyannaz sandstone of the Swiss Alps and equivalent sandstones in France and Italy, Journal of Sedimentary Petrology 49, 463-481

STAPLIN, F.L. (1969): Sedimentary organic matter, organic metamorphism and oil and gas occurrence, Can. Petr. Geol. Bull. 17, 47-66

STEGENA, L. (1967): On the formation of the Hungarian basin (in Hungarian), Földtani Közlöny 97, 278-285

STEGENA, L., F. HORVÁTH, J.G. SCLATER & L. ROYDEN (1981): Determination of paleotemperature by vitrinite reflectance data, Earth Evol. Sci. 3-4, 292-300

STEGENA, L. & P. DÖVÉNYI (1983): Procedure to correct measured temperatures and temperature gradients for geological effects, Zbl. Geol. Palaont., Teil I, 1-2, 25-34

STRASSER, B. & R. WOLTERS (1963): Gesteinsmechanische Untersuchungen an Proben aus der Bohrung Münsterland 1, Fortschr. Geol. Rheinland u. Westf. 11, 419-446

STREET, F.A. & A.T. GROVE (1976): Environmental and climatic implications of late Quaternary lake-level fluctuations in Africa, Nature 261, 385-390

STREET, F.A. & A.T. GROVE (1979): Global maps of lake-level fluctuations since 30,000 B.P., Quaternary Research 12, 83-118

SUGGATE, R.P. (1982): Low-rank sequences and scales of organic metamorphism, Journal of Petroleum Geology 4, 377-392

SUZUOKI, T. & P. EPSTEIN (1976): Hydrogen isotope fractionation between OH-bearing minerals and water, Geochim. et Cosmochim. Acta 40, 1229-1240

SZÁDECZKY-KARDOSS, E. (1971): The Carpatho-Dinarid area from the point of view of the new global tectonics (in Hungarian), Geonómia és Bányászat 4, 3-89

SZÁDECZKY-KARDOSS, E. (1973): Die chemische Zusammensetzung der natürlichen Kohlenwasserstoffe und die geologische Struktur, VII. Geochem. Konf. U.G.J., in: OGIL, 154-184, Budapest

TAGUCHI, S., T. FUJINO & M. HAYASHI (1979): Homogenization temperature measurements of fluid inclusions in quartz and anhydrite from the Hatchobarn geothermal field, Japan, and its applications for geothermal development, Geothermal Resources Council, Transactions 3, 705-708

TAN, Li-ping (1965): The metamorphism of Taiwan Miocene coals, Taiwan Geological Survey Bulletin 16, 44 p.

TAYLOR, J.C.M. (1983); Bit metamorphism can change character of cuttings, Oil and Gas Journal 81, 107-112

TEICHMÜLLER, M. (1970): Bestimmung des Inkohlungsgrades von kohligen Einschlüssen in Sedimenten des Oberrheingrabens - ein Hilfsmittel bei der Klärung geothermischer Fragen, in: J.H. ILLIES & St. MÜLLER (eds.), Graben Problems, 124-142, Stuttgart (Schweizerbart)

TEICHMÜLLER, M. (1971): Application of coal-petrographic methods in petroleum and natural gas prospecting, Erdöl u. Kohle 21, 69-76

TEICHMÜLLER, M. (1971): Das Inkohlungsprofil des flözführenden Oberkarbons der Bohrung Isselburg 3 nordwestlich Wesel, Geol. Mitt., 11

TEICHMÜLLER, M. (1979): Die Diagenese der kohligen Substanzen in den Gesteinen des Tertiärs und Mesozoikums des mittleren Oberrhein-Grabens, Fortschr. Geol. Rheinland u. Westf. 27, 19-49

TEICHMÜLLER, M. (1982): Rank determination on sedimentary rocks other than coal, in: E. STACH, M.Th. MACKOWSKY, M. TEICHMÜLLER, G.H. TAYLOR, D. CHANDRA & R. TEICHMÜLLER (eds.), Textbook of Coal Petrology, 363-371, Stuttgart (Borntraeger)

TEICHMÜLLER, R. (1973): Die paläogeographisch-fazielle und tektonische Entwicklung eines Kohlenbeckens am Beispiel des Ruhrkarbons, Z. dt. geol. Ges. 124, 149-165

TEICHMÜLLER, M. & R. TEICHMÜLLER (1949): Inkohlungsfragen im Ruhrkarbon, Z. dt. geol. Ges. 99, 40-77

TEICHMÜLLER, M. & R. TEICHMÜLLER (1958): Inkohlungsuntersuchungen und ihre Nutzanwendung, Geol. en Mijnb. 20, 41-66

TEICHMÜLLER, M. & R. TEICHMÜLLER (1968): Geological aspects of coal metamorphism, in: D.G. MURCHISON & T.S. WESTOLL (eds.), Coal and Coal Bearing Strata, 233-267, Edinburgh (Oliver and Boyd)

TEICHMÜLLER, M. & R. TEICHMÜLLER (1971): Inkohlung im Rhein-Ruhr-Revier, Fortschr. Geol. Rheinland u. Westf. 19, 47-56

TEICHMÜLLER, M. & R. TEICHMÜLLER (1975): Inkohlungsuntersuchungen in der Molasse des Alpenvorlandes, Geol. Bavarica 73, 123-142

TEICHMÜLLER, M. & R. TEICHMÜLLER (1978): Coalification studies in the Alps, in: H. CLOSS, D. RODER & K. SCHMIDT (eds.), Alps, Apennines, Hellenides, Scient. Report 38, 49-55

TEICHMÜLLER, M. & R. TEICHMÜLLER (1979a): Zur geothermischen Geschichte des Oberrhein-Grabens. Zusammenfassung und Auswertung eines Symposiums, Fortschr. Geol. Rheinland u. Westf. 27, 109-120

TEICHMÜLLER, M. & R. TEICHMÜLLER (1979b): Diagenesis of coal (coalification), in: G. LARSEN & G.V. CHILINGAR (eds.), Diagenesis in sediments and sedimentary rocks, Developments in sedimentology 25A, 207-245, Amsterdam (Elsevier)

TEICHMÜLLER, M. & R. TEICHMÜLLER (1981): The significance of coalification studies to geology - a review, Bull. Centre Rech. Expl. - Prod. Elf-Aquitaine 5, 2, 491-534

TEICHMÜLLER, M. (1982): Application of coal petrological methods in geology including oil and natural gas prospecting, in: E. STACH, M.Th. MACKOWSKY, M. TEICHMÜLLER, G.H. TAYLOR, D. CHANDRA & R. TEICHMÜLLER (eds.), Textbook of Coal Petrology, 381-413, Berlin-Stuttgart (Borntraeger)

TEICHMÜLLER, M., R. TEICHMÜLLER & V. LORENZ (1983): Inkohlung und Inkohlungsgradienten in der Saar-Nahe-Senke, Z. dt. geol. Ges. 134, 153-210

TILLMAN, J.E. & H.L. BARNES (1983): Deciphering fracturing and fluid migration histories in northern Appalachian basin, A.A.P.G. Bull. 67, 4, 692-705

TING, F.T.C. (1978): Petrographic techniques in coal analysis, in: C. KARR jr. (ed.), Analytic Methods for Coal and Coal Products, 3-26, New York (Academic Press)

TISSOT, B. (1969: Premières donnéas sur les mécanismes et la cinétique de la formation du pétrole dans les sédiments. Simulation d'un schéma réactionell sur ordinateur, Inst. franç. Pêtrole Rev. 24, 470-501

TISSOT, B.P., J.F. BARD & J. ESPITALIE (1980): Principal factors controlling the timing of petroleum generation, in Facts and principles of world petroleum occurrence, Canadian Soc. Petroleum Geologists Mem. 6, 143-152

TISSOT, B., G. DEROO & J. ESPITALIE (1975): Etude comparee de l'epoque de formation et d'expulsion du petrole dans diverses provinces geologiques, Ninth World Petr. Congr. Proc. 2, 159-169

TISSOT, B. & J. ESPITALIE (1975): L'evolution thermique de la materie organique des sediments: applications d'une simulation mathematique, Inst. Français Petr. Rev. 30, 743-777

TISSOT, B.P. & D.H. WELTE (1984): Petroleum Formation and Occurrence, second edition, New York, 699 p. (Springer)

TRICART, J. (1975): Existence de périodes sèches au Quaternaire en Amazonie et dans les régions voisines, Revue Géographie Dynamique 23, 145-158

TROFIMUK, A.A., N.V. TSERSKY, V.P. TZAREV, E.M. GALIMOV, O.L. KUZNETZOV, T.I. SOROKO & V.G. TSAKHMATSEV (1983): Seismotectonic processes - factors of the maturation of organic matter of sediments, Dokl. AN URSS 271, 6, 1460-1464

TUGARINOV, A.I., & V.B. NAUMOV (1970): Dependence of the decrepitation temperature of minerals on the composition of their gas-liquid inclusions and hardness, Dokl. Acad. Sci. USSR 195, 112-114

TURCOTTE, D.L. (1980): On the thermal evolution of the Earth, Earth Planet. Sci. Lett. 48, 53-58

UREY, H. (1947): The thermodynamic properties of isotopic substances, J. Chem. Soc., 562-682

VALLENTYNE, J.R. (1964): Biogeochemistry of organic matter II, Geochim. et Cosmochim. Acta 28, 157-188

VALLEY, J.W. & J.R. O'NEIL (1981): $^{13}C/^{12}C$ exchange between calcite and graphite: a possible thermometer in Greenville marbles, Geochim. et Cosmochim. Acta 45, 411-419

VAN DE KAMP, P.C. (1976): Inorganic and organic metamorphism in siliciclastic rocks (abst.), A.A.P.G. Bull. 60, 279

VAN KREVELEN, D.W. (1967): Geochemistry of Coal, in: I.A. BREGER (ed.), Organic Geochemistry, 183-247 (Pergamon Press)

VEIT, E. (1963): Der Bau der südlichen Molasse Oberbayerns aufgrund der Deutung seismischer Profile, Bull. Ver. Schweiz. Petrol. Geol. u. Ing. 30 (78), 15-52

VELEV, V.H., S.P. VETSEVA & G.D. SHISHKOV (1979): Catagenesis of organic matter and clay sediments (in Bulgarian), Ann. Univ. Sofia, Geology 73, 19-64

VETÖ, I. (1980): An examination of the timing of catagenesis of organic matter using three published methods, in: A.G. DOUGLAS & J.R. MAXWELL (eds.), Adv. Org. Geoch., 163-167

VETÖ, I., P. DÖVÉNYI & I. KONCZ (1984): Critical comparison of some methods for the geothermal reconstruction on the basis of vitrinite reflectance, Acta Geodaet., Geophys. et Mountanist. Hung. 19, 1-2, 161-171

VISSER, W. (1982): Maximum diagenetic temperature in a petroleum source rock from Venezuela by fluid inclusion thermometry, Chem. Geol. 37, 95-101

VÖLGYI, L. (1977): The role of geothermal conditions and hydrocarbon prognostic (in Hungarian), Acta Geol. Acad. Sci. Hung. 21, 143-167

WADA, H. & K. SUZUKI (1983): Carbon isotopic thermometry calibrated by dolomite-calcite solvus temperatures, Geochim. et Cosmochim. Acta 47, 697-706

WAGNER, G.A. (1968): Fission track dating of apatites, Earth Planet. Sci. Lett. 4, 411-415

WAGNER, G.A., D.S. MILLER & E. JÄGER (1979): Fission track ages on apatite of Bergell rocks from Central Alps and Bergell boulders in Oligocene sediments, Earth Planet. Sci. Lett. 45, 355-360

WAGNER, G.A. & G.M. REIMER (1972): Fission track tectonics: the tectonic interpretation of fission track apatite ages, Earth Planet. Sci. Lett. 14, 263-268

WAGNER, G.A., G.M. REIMER & E. JÄGER (1977): Cooling ages derived by apatite fission-track, mica Rb-Sr and K-Ar dating: the uplift and cooling history of the Central Alps, Mem. Ist. Geol. Min. Univ. Padova, Vol. XXX

WANG JI-AN, WANG JI-YANG, WANG JUN, HUANG GE-SHAN, YAN SHUZHEN, LU XIN-WEN (1983): Geothermal studies in oil field district of north China, MS presented at IUGG Gen. Ass. Hamburg

WAPLES, D.W. (1980): Time and temperature in petroleum formation: Application of Lopatin's method to petroleum exploration, Bull. Amer. Assoc. Petrol. Geol. 64, 916-929

WAPLES, D.W. (1981): Organic Geochemistry for Exploration Geologists, Minneapolis, 151 p. (Burgess Publishing Co.)

WAPLES, D.W. (1983): Physical-chemical models for oil generation, Colorado School of Mines Quarterly 78, 4

WAPLES, D.W. (1984): Thermal models for oil generation, in: J. BROOKS & D. WELTE (eds.), Advances in Petroleum Geology, Vol. 1, 8-67

WEAVER, C.E. (1979): Geothermal alteration of clay minerals and shales: diagenesis, Office of Nuclear Waste Isolation Technical Report 21, 176

WEBB, T. (1985): A global paleoclimatic data base for 6000 B.P., U.S. Department of Commerce Technical Report TRO 8, Washington D.C., Office of Energy Research

WEBER, K. & H.-J. BEHR (1983): Geodynamic interpretation of the Mid-European Variscides, in: H. MARTIN & F.W. EDER (eds.), Intracontinental fold belts, 427-469, Berlin-Heidelberg (Springer)

WEBER, V. (1984): Vorausberechnung der Gebirgstemperaturen im Ruhrkarbon in Abhängigkeit vom Schichtenaufbau, Fortschr. Geol. Rheinland u. Westf. 32, 283-296

WEGENHAUPT, H. (1962): Zur Petrographie und Geochemie des höheren Westfal A von Westerholt, Fortschr. Geol. Rheinland u. Westf. 3, 2, 445-496

WEHNER, H., H. DAMBERGER, D. LEYTHAEUSER & D.H. WELTE (1976): Organisch-geochemische Untersuchungen an Kohlen, Gesteinen und Restgasen aus der Bohrung Saar 1, Geol. Jb. A27, 455-488

WEISS, A. & G. ROLOFF (1965): Über die Rolle glimmerartiger Schichtsilikate bei der Entstehung von Erdöl und Erdöllagerstätten, Diss. Heidelberg

WELTE, D.H. & M.A. YUKLER (1981): Petroleum origin and accumulation in basin evolution - a quantitative model, A.A.P.G. Bull. 65, 1387-1396

WERNER, D. (1980): Probleme der Geothermik im Bereich der Schweizer Zentralalpen, Eclogae Geol. Helv. 73/2, 513-525

WERNER, D. (1981): A geothermic method for the reconstruction of the uplift history of a mountain range, applied to the Central Alps, Geol. Rdsch. 70, 296-301

WERNER, D. (1985): Geothermal problems in mountain ranges (Alps), Tectonophysics, in press

WERNER, D. & E. KISSLING (1985): Gravity anomalies and dynamics of the Swiss Alps, Tectonophysics 117, in press

WHITE, D.E. (1965): Saline waters of sedimentary rocks, in: Fluids in Subsurface Environments Symp., Mem. Am. Petr. Geol. 4, 342-366

WHITEHEAD, E.V. (1983): Geochemistry of natural products in petroleum prospecting, in: G.B. CRUMP (ed.), Petroanalysis '81: Advances in analytical chemistry in the petroleum industry 1975-1982, 31-75 (Wiley)

WOLF, M. (1961): Sporenstratigraphische Untersuchungen in der gefalteten Molasse der Murnauer Mulde (Oberbayern), Geologica Bavarica 46, 53-92

WOLF, M. (1963): Sporenstratigraphische Untersuchungen im Randcenoman Oberbayerns, N. Jb. Geol. Palaont. Mh., 337-354

WRIGHT, N.J.R. (1980): Time, temperature, and organic maturation - the evolution of rank within a sedimentary pile, Journal of Petroleum Geology 2, 411-425

YORATH, C.J. & R.D. HINDMAN (1983): Subsidence and thermal history of Queen Charlotte basin, Can. J. Earth Sci. 20, 1, 135-159

ZAGORUCHENKO, V.A. & A.M. ZHURAVLEV (1970): Thermophysical properties of gaseous and liquid methane, Israel Progr. Sci. Transl., Jerusalem (russ. ed.: 1969)

ZAUN, P. & G.A. WAGNER (1984): Ein Stabilitätsprofil von Spaltspuren im Zirkon. Neue Wege zur Ermittlung der thermischen Geschichte der Erdkruste, Sitzungsberichte 14. Sitzung FKPE-Arbeitsgruppe "Ermittlung der Temperaturverteilung im Erdinnern", 8-9, Clausthal-Z.

ZEN, E-an & A.B. THOMPSON (1974): Low grade regional metamorphism: Mineral equilibrium relations, Annual Reviews of Earth and Planetary Science 2, 179-212

ZIELINSKI, G.W. & P.M. BRUCHHAUSEN (1983): Shallow temperatures and thermal regime in hydrocarbon province of Tierra del Fuego, A.A.P.G. Bull. 67, 166-177

ZIMMERLE, W. (1976): Petrographische Beschreibung und Deutung der erbohrten Schichten (in der Bohrung Saar 1), Geol. Jb. A27, 91-305

ZINKERNAGEL, U. (1978): Cathodoluminescence of quartz and its application to sandstone petrology, Contr. Sediment. 8, 1-69

ZWART, H.J. (1967): The duality of orogenic belts, Geol. Mijnb. 46, 283-309

ZWART, H.J. (1976): Regional metamorphism in the Variscan orogeny of Europe, Franz Kossmat Symposium, Nova Acta Leop., N.F. No. 224, 45, 361-367

ZUBENKO, V.G., S.D. PUSTIL'NIKOVA, N.N. ABRYUTINA & Al.A. PETROV (1980): Neftyanye monoaromaticheskiye uglevodorodi steroidnovo tipa (Petroleum monoaromatic hydrocarbons of steroid type), Neftekhimiya 20, 490-497

ZUBENKO, V.G., N.S. VOROB'EVA, Z.K. ZEMSKOVA, T. PENK & Al.A. PETROV (1981): O raynobesii tsis-i trans-izomerov oktagidrofenantrenov - strukturnykh fragmentov monoaromaticheskikh steranov (Equilibrium of cis- and trans-isomers of octahydrophenanthrenes as structural fragments of monoaromatic steroids), Neftekhimiya 21, 323-328

SUBJECT INDEX

absolute time calculations 146
absolute time method 138,144,146,147,149, 150
activation energy 14,121
activation enthalpy 154,155,159
Adriatic Sea 71,73
alginite 106
alkane 125,127
Alpidic Foredeeps 53,59,67,77
Alpine lithosphere 186
Alpine - Mediterranean region 174
Alpine molasse 68,71
Alpine region 190
Alps 24,32,38,39,53,57,59,67,68,70,71,77, 78,185
ambient water 30,31
amphibolite facies 49
anchimetamorphism 57
anisotropy 88
annealing temperature 38
anthracite 5,9,158
anti conformations 126
Anzing-3 68,70,71
apatite 185,188
Apennines 53,59,71-73,75-78
Apenninic foredeep 71
Appalachian Basin 26
Apuan Alps 72
argillaceous sediments 31
aromatization 54,119,121,122,128,129,131- 133,135,137,142,146-151,153,157,160-163, 166,167,170-172
aromatic rings 128
Arrhenius 136,145
Arrhenius equation 9,20,37
Arrhenius-plot 36,142,143,144,146-149
asphaltene 8,129,130
astronomical changes 41
Asturian folding 59
Athabasca 25
Atlas of Oceans 108
Atreo 73
Australia 95,97-103
Austroalpine nappe 67
authigenic minerals 31,34,90,195,199
average deviance 111

bacterio-hopanetetrol 122,126,127
Baoding 202,204
basic-temperature 28
basin formation 180
Bass Basin 98-100,102
Bavarian Alps 68

Bergell 39
biological markers 121,122,126,129,130, 137,150,154,158
biotite 185
bireflectance 88
bishomohopane 126
bitumen 8,13,130,131,160
bituminous coal 9,67,70,72
bituminous products 7
blocking temperature 185,188,192,193
B model 112,114
Bohai 196
Bohemian Mass 77
Boltzmann distribution 133
Bone Spring 92
Boomi-1 99,101,103
Bostick-method 105,107,108,111,120,158, 178,179,181-183
Börzsöny 180
Braunkohle 67
Brazoria County 29
Broadlands 27
brown coal 7,8
Buntebarth method 98,100-103,120
burial history 10,12,17,20,53,55,57,64, 79,82,84,89,105,108,109,112,188
burial rate 96
buried hill 196

calc-alkaline volcanic activity 180,183
calcite-graphite fractionation 51
calcite-water oxygen isotope 92
California 71
Canaxian 203
Cardium sandstone 23,24
Carpathian arc 130
Carpathian basin 130
Carpathian Mts. 71
Carrara 72
catagenesis 123,129
catalytic effect 124,126,133
catastrophe 161,164,170,171
cation exchange 47
cellulose 54
Cenozoic reservoirs 196
centivalli line 193
central Alps 185-188,190,191
central Prairies Basin 17,18
Cerro Prieto 29,32,33,38,90
changes in precipitation 42
China 195-197
chiral centers 123,125,126,128,129
chloroform extract 131

cholestane 122,124,135
circulation of water 110
clay minerals 31,34,90,195
CLIMAP 41
climatic response 42
closing temperature 38
coal deposits 6
coalification 6,9,53,59,61,62,67,68,72,73,
 75,77,96,103,106,107,111,120,121,137,159,
 160,162
coalification gradient 18,53,62,62,68,69,
 71,77,78,200
coalification studies 67,71
coalification temperature 57
coal macerals 9
coal rank 16,17,32,91
cogenetic minerals 30,31
Colline Metallifere 73
collision frequency 136
color alteration index 36,37
compaction 11,12,23
concordant temperature method 46
conodonts 35-37
continuity equation 187
continuously subsiding basins 111,112
convective heat transport 175
conversion 156-159,161-168,170
cooking time 53,54
cooling basin 155
cooling history 38,39,192
cooling rate 186
Correction for compaction 109
correlation coefficient 83
crustal overthrusting 185
crustal thickening 187
crystallization 47

decomposition of plants 7
decrepitate 25
degree of coalification 8,9,53,54,70,71,73,
 95,124
degree of diagenesis 9,32
degree of transformation 156
$\delta^{18}O$ 30,92
delta-δ-value 29,30
diagenesis 7,9,31,79,81,84,86,121
diagenesis of montmorillonite 31
diagenesis of organic matter 6,32
diagenesis of sedimentary rocks 34
diagenetic dehydration 128
diagenetic history 79,88
diaromatic compounds 128,129
diastereomers 125
differential method 139,142,143,145,148-
 150
diffusion 48
dyke 175,177,179-183

effective heating time 15,54,55,63,81,82,
 107,137,140-143,145,147-150
Egling 68

Eiweiler-Vogelsborn 98,99,101
Elba 71,73,76
electron spin resonance 15
ellipticity 42
enantiomers 125,128
energy barrier 133
enthalpy 119,133,134,136,139,142,143,
 158,159,161,164,167,168,170-172
entropy 119,134,136,142,143,146
epimeric mixtures 123
epimerization 123,127,133
epi-zone 66
equilibrium 136,137
equilibrium constant 46,134-136,142,145,
 156
erosion rate 109
error propagation 170
ethylcholestanes 123,125
eucaryotes 126
exchange reactions 52
Exmouth Plateau 102
expandable layers 32,33
extensional tectonics 110

Falkland Plateau 18
fast sedimentation 110
Finlandia Vein 50,51
first-order reaction 9,132,133,147
fission track 92,120,185,188
fluid inclusions 23,25,26,30,32,50,92
fluid inclusion thermometry 22,91
Flysch 67,71
formation temperature 47
fractionation 27,49,51
fractionation factor 29,30,46
fragmentogram 124,125,127,129
free radicals 15,129
freezing point 23
freezing temperatures 25
frictional heating 186,187,189,190,192
frictional zone 193
FRG 99,100-102,114
fusunite 13

gammacerane 127
gas chromatography 123,124
gas coal 8
gauche-conformations 126
GC-MS analyses 127,129,134
geochemical thermometry 27
geosteroids 129
geothermal systems 52
geothermometry 51
German coal classification 67
Germany 19,38,64,95,98,103
Girondelle Seam 62
glacial-interglacial 41-43
Gondwana coal seams 107
Gotthard 185,186,188-191
granulite facies 49
graphitization 90

gravity sliding 190
Great Hungarian Plain 130,131
Green River Basin 91
greenschist assemblages 49
greenschist metamorphism 78,81
Grosseto 72
groundwater 120
Gulf Coast 121
Gunnedāh Basin 102

Harthausen-1 70
Hatchobaru 26,27
Hausham 68
heating duration 54,80-84,86,90-92
heating experiments 135
heating rates 153,161,163-165,167,168,171, 172
heat transport 187
Hebei 195,200,202,203
Helveticum 67
hematite 49
high heating rate 169
highly matured basins 112
H model 112,114
Hod-I 130,131,142,150
Hokkaido 34
Holocene 43
homogenization temperature 22-27,91,92
homohopane 126,127
Hood's method 105,107,111,120,137,158,201, 202
hopane 19,20,21,119,126,127,131,133,137, 142,145-147,150,151,170,171
hopanoid 122,126,127,129
huminite 8
Hungary 181
hydride iron 126
hydrocarbon generation 82,174
hydrogen 51
hydrogen radical 126
hydrothermal alteration 32,33
hydrothermal metamorphism 90
hydrothermal mineral 92
hydrothermal systems 90

igneous masses 178,180,182,183
illite crystallinity 32
illite diagenesis 68
illite/smectite 91
Imperial Valley 26
inclusions 24,25
inertinite 88
intracontinental basins 110
intrusion 176,177,183
isomerization 119,121-124,126-129,131-133, 135-137,142-147,150-151,161-163,167,168, 170
isomers 124,127,129,136
isorank lines 62
isoreflectance line 84,183
isotope equilibrium 45,46-48
isotope exchange 27,28,31,45-52

isotope fractionation 30,45-47,51,52
isotope geothermometers 45,51
isotope partitioning 45
isotope thermometry 29,52
isotopic composition 47-49,51
isotopic temperatures 30,49,50,52
Isselburg 66

Japan 32,34
Jupiter-1 98,99,101
K/Ar-method 78
Karweil method 119,200,201
katabatic winds 42
Katharina 63
kerobitumen 9
kerogen 8,14,15,121,129-131,158,159,173, 202
Kerogen metamorphism 90
Krefeld pluton 63
kinematic model 188
kinetic equation 132
kinetic parameters 153,154,160
Kupferschiefer 62

laboratory bias 87
Langfang-Guan 202,204
Larderello 73
latent heat 175
lava flows 182
level of maturity 106,178
level of organic metamorphism 15
Liaohe 18,195,197-200,202-204
lignin 54
Liguride-Luretta-Sporno nappe 72
liquid window 202,203,204
little ice age 43
L models 112,114
Lopatin method 10,105,107,111,119,158,173, 176,178,181-183,201
low heating rate 169,172
Lower Saxony Basin 19
low pressure metamorphism 77

magma 175
magmatic activity 174,179
magmatic bodies 183
magmatic heat 181,182
mantle diapir 77
mantle heat flow 110
marker reactions 20,119,132
Martina 171
mass spectrometer 124
mass transfer 132
mathematical model 155
Mathilde 71
Mátra Mts. 181
maturation 16,86,120,173,174
maturation of organic matter 14,15,105, 173,176,178,179,181-183
maturity 18,105,106,112
maturity changes 179
maturity increase 182

maturity parameter 10
maximum burial temperature 54,79,80
maximum geothermometer 79,92
maximum heating 114
maximum paleotemperature 15
maximum temperature 79-84,86,89,91,92,107, 108,131,137,179,200,201
measurement errors 136
Mediterranean Sea 43
Metamorphic mineral assemblages 90,92
metamorphism 48,49,51-53,66,72,73,77,78, 80,89-91,188,191
method of absolute time 143
methyl groups 127
microscope photometer 54
microthermometry 51
Middleton method 98,100-104
Miesbach 167
Minas Gerais 49
mineral metamorphism 90
mineral pairs 45-47,50-52
mineral synthesis 47
mineral zones 90
models 105
Mohorovičič discontinuity 68,73
molasse 67-71
Molasse basin 18
monoaromatics 128,129,132-134,147
monoaromatic steroids 151
Monterey 37
Monte Rosa 192,193
montmorillonite-illite-transformation 31, 32
Moscow lignites 121
Mundry's formula 175
Münsterland 63,64,66,86,112,114
Murrelet 115
muscovite 185

Nahe Basin 77
Na-K-Ca thermometer 28
naphthenes 160
Newtonian fluid 187
New Zealand 26,107
Niigata basin 35
Nonantola 73
Nord coal basin 107
Nordlicht-Ost 1 98,99,101
North Atlantic 18
North China 195,196-199,201-204
Northern Alps 67
Northern Appennines 71
North Hungary 173,174,180,181,183
Northern Shandong 195
North Sea 20,121
North Shanxi Basin 197
Nova Scotia 18
nuclear events 185
numerical integration 139

^{17}O 48
^{18}O 29,31
Odenwald 38
OH-bearing minerals 51
oil formation 19
oil generation 10,13,130
oil-generation window 173,180,183
oil geochemistry 13
oil window 112
operator bias 87
optical reflectivity 8,54
orbital geometry 41
organic debris 15
organic matter 6
organic maturation 16
organic metamorphism 119
Oswego fault system 26
overthrusting 190
oxygen 52
oxygen isotope 47,48,92
oxygen isotopic equilibrium 31

paleogeothermal gradients 18,62,77,78,95, 96,102
Pannonian 119,169
Pannonian Basin 20,77,130,131,137,139,141, 142,151,161,162,163,164,165,166,168,172, 183
parameters reaction 167
Paris Basin 123
peat 8,9
peatification 7
Pecten-1A 98-100
Penninic nappes 68
Pergamon World Atlas 108
Permian Basin 92
petroleum migration 31
Petrov's isomerate 135
phengite 185
plutonic body 175,176
point-source solution 174
Ponte dell'Olio 71,72
Po Plain 71-73,76
Pratomagno 73
precession 42
pre-exponential factors 119,121,134,136, 142,146,153-155,157-159,163,164,166,167, 169,170-172
pressure correction 22-24
pressure in oil generation 14
primary inclusions 22,26
primitive function 139
pristane 126
procaryotes 126
proportionality constant 136
Providencia mine 50,51
pyrolysis 14,130,160

quartz 49

R_o 8,11-13,15,16,106-109,111,112,114,121, 173,180-182,201
racemization 42
radiogenic heat sources 186,189,190
radiometric dating 38
rank 13,14,53-57,62-64,70-72,79,80,82,84, 87-91,114
rank gradient 53,55,59,62-65
rank of coalification 6,8-10
rank range 65
rate constant 135-139,142,149,154
rate equation 155
rate law 135
rate of heating 157-159,165,166
rate of mass transfer 148
rate of subsidence 159
rate parameters 130,132,137,138,140-142, 149-151,157
reaction integral 138-140
reactions kinetics 119,153
reaction order 132
reaction rate 130,132,133,154
recrystallization 47,49
rectangular prism 174-176,180,181
reflectance 13
regression analysis 79,80,82,83,92
relative deviance 111
resin 8,160
resinite 13
retention temperature 39
retrograde isotope exchange 48,52
Rhenish Variscan Mts. 78
Rhenohercynicum 67
Rhine Basin 59,62,63,77,78
Rhine Graben 55,57-59,62,66,70,77,102
rhyolitic tuffs 183
R-isomers 132,134
R_m 8,16,17,53,55,64,66,70,73,76,79-82,84, 86-93,96,98,103,106
R_{max} 8,64,72
R_{min} 64
rotational freedom 136
Rotondo 190
Ruhr 59,61,62,66,70,71,77,78,121
Ruhr Basin 18,53,59,62-65,67,102,114

Saar 57,60,63-66
Saar Basin 19,59
Saar-Nahe Basin 57,60,77,102
Salton Sea 27,30
Sandhausen-1 70,98-100,103
schistosity 49,50
secondary inclusions 22
second-order reactions 132
sedimentary basins 173
sedimentary history 11,12,23,24
semi-anthracite 8
Shandong 196,202,203
Shijiazhuang-Jinan line 202,203
Sichuan Basin 197
silica geothermometer 92

silica minerals 34
silica polymorphs 34
silicate-pair fractionations 48
Simplon 192,193
SiO_2-thermometer 28
S-isomeres 132,134
smectite/illite 31
sodium-calcium-potassium geothermometer 92
Southern Alps 191,193
Southern Continental Margin 102
Southern Island 107
spectral fluorescence 9
sphalerite-galena pair 50,51
sporinite 9
stable isotopes 29
Staffelsee 68
statistical mechanical theory 46
Steinkohle 67
sterane 19-21,122,123-128,131-133,136, 142-144,146,150,151,161,163,168,170,171
stereochemistry 122,128,129
stereoisomeres 125
steric hindrance 125
steroids 19,20,119,122,129,130,137,142,147
sterol 122,126,128
stratovolcano 180-182
stretching 20,21,120
Subalpidic foredeeps 77
sub-Apenninic sediments 71
sub-bituminous 67
subduction 67,68,70,71,78
subsidence rate 73,96,98,102,103,110
subsidence/uplift history 62,70,96,97,103
Subvariscan foredeeps 53,59,67,77,78
Subvariscan Molasse 71
sulfide geothermometry 50
sulfide pair fractionations 50
sulfur isotope fractionations 51
Swabian Alb 19,102
swamp types 6
Swiss Helveticum 66

Taihang 199,202
tectonic lineament 193
tectonic models 120
Tegernsee 68
telemagmatic effect 173,183
temperature at the sea floor 42
temperature calibrations 45,46
temperature coefficients 50
temperature corrections 88
temperature history 52
temperature-time-index 12,14,32
temperature-time-models 90
temperature-time rank models 80,89,92
temperature-time-reflectance 120
tensional tectonics 73
terpanes 130
thermal alteration 14,15,121
thermal catastrophe 155,156,159,164

thermal event 49,179
thermal history 10,16,21,32,37,38,88,105-107
thermal jump 158,162
thermal maturity 11,15,79-82,84-86,89,90,92,180
thermal memory 185
thermal plume 32
three isotope approach 47
Ticino 193
tilt 42
time-temperature history 89
Time Temperature Index 10,11,33,86,108,121,176
torbanitic oil shales 106
transformation of minerals 31
translational freedom 136
transmission coefficient 136
transmission electron microscopy 14
triaromatics 128,129,132-134,147
triaromatization 126
triterpane 127
triterpenoids 126
TTI 12,14,32
Tuscany 73
Tyrrhenian Sea 72,73

United States 179
uplift history 185,188,189,191,192
uplift rate 186
Upper Rhine Graben 19,32
Urach-3 98-100

Variscan 53,59,64,77
Variscicum 77

Venezuela 26
Versmold 66
vibrational frequencies 47
vibrational period 133
vibration-rotation 133
vitrinite 7-9,13,54,87,88,107,120,153,162
vitrinite populations 87
vitrinite reflectivity 8,11-19,37,53-60,62-64,66,67,71,72,76,79-81,83,87,88,91,92,96-101,103-108,111-115,119,120,171,173,174,176,178-182,195,199,200
vitrinite-type reactions 162-164,166-168,172
vitrites 56
volcanic activity 173,179,182
volcanic body 175
volcanic dykes 174,178
volcanic effect 110
volcanic heat 178
volcanism 196
Vorderriss 68

Wagon Wheel-1 91
Wairakei 52
Waples method 105,107,108,111,120,176,181,182,201
water migration 28,29
Westerholt 63
W method 112,114
woody kerogen 8

Yanshan 199,202
Yintze tectonic cycle 195

zeolites 32,34,90